CW01371815

Tanks in the Great War, 1914-18

Big Willie, Mother of Centipede, Original Mark I Tank.

Tanks in the Great War, 1914-18

The Development of Armoured Vehicles and Warfare

ILLUSTRATED
WITH 40 PHOTOGRAPHS, MAPS, AND DIAGRAMS

J. F. C. Fuller

LEONAUR

Tanks in the Great War, 1914-18
The Development of Armoured Vehicles and Warfare
by J. F. C. Fuller

ILLUSTRATED WITH 40 PHOTOGRAPHS, MAPS, AND DIAGRAMS

First published under the title
Tanks in the Great War

Leonaur is an imprint of Oakpast Ltd
Copyright in this form © 2019 Oakpast Ltd

ISBN: 978-1-78282-886-0 (hardcover)
ISBN: 978-1-78282-887-7 (softcover)

http://www.leonaur.com

Publisher's Notes

The views expressed in this book are not necessarily those of the publisher.

Contents

Dedications	7
Introduction	7
The Origins of the Tank	15
The Invention of the Landship	27
Mechanical Characteristics of Tanks	43
The Mark I Tank and Its Tactics	54
The Battles of the Somme and Ancre	59
The Growth of the Tank Corps Organisation	63
Tank "Esprit De Corps"	69
Tank Tactics	73
The Battle of Arras	79
Tank Battle Records	87
The Second Battle of Gaza	93
Staff Work and Battle Preparation	99
The Battle of Messines	102
A Tactical Appreciation	106
The Third Battle of Ypres	109
Tank Mechanical Engineering	117
The Third Battle of Gaza	121

Origins of the Battle of Cambrai	125
The Battle of Cambrai	129
An Infantry Appreciation of Tanks	141
The Tank Corps Training Centre	145
The Tank Supply Companies	151
The Second Battle of the Somme	156
Tank Signalling Organisation	161
The French Tank Corps	166
Preparations for the Great Offensive	180
The Battles of Hamel and Moreuil	184
German Tank Operations	193
The Battle of Amiens	197
The Fight of a Whippet Tank	210
German Appreciation of British Tanks	215
Aeroplane Co-Operation with Tanks	220
The Battle of Bapaume and the Second Battle of Arras	227
German Anti-Tank Tactics	235
The Battles of Epehy and Cambrai—St. Quentin	239
The U.S.A. Tank Corps	248
The Battles of the Selle and Maubeuge	253
The 17th Tank Armoured Car Battalion	257
A Retrospect of What Tanks Have Accomplished	263
A Forecast of What Tanks May Do	271

Dedications

1

I dedicate this book to the modern military scientists, that small company of gentlemen who, imbued with a great idea, were willing to set all personal interest aside in order to design a machine destined to revolutionise the science of war.

2

I dedicate this book to the modern armourers of the British factories, those men and women whose untiring patriotism and indomitable endurance in the workshops produced a weapon whereby the lives of many of their comrades were saved.

3

I dedicate this book to the modern knights in armour, the fighting crews of the Tank Corps; those officers, non-commissioned officers and men, who, through their own high courage and noble determination on the battlefield, maintained liberty and accomplished victory.

Introduction

The following work is the story of a great and unique adventure as heroic as the exploits of the Argonauts of old, and, though the time perhaps has not yet arrived wherein to judge the part played by tanks in the Great War, I feel that, whatever may be the insight and judgment of the eventual historian of the British Tank Corps, he will probably lack that essential ingredient of all true history—the witnessing of the events concerning which he relates.

I, the writer of this book, first set eyes on a tank towards the end of August 1916. At this time, I little thought that I should eventually be honoured by becoming the Chief General Staff Officer of the Tank Corps, for a period extending from December 1916 to August 1918.

The time spent during this long connection with the greatest military invention of the Great War, it is hoped, has not been altogether wasted, and the story here set forth represents my appreciation of having been selected to fill so intensely interesting an appointment.

Besides having witnessed and partaken in many of the events related, those who have assisted me in this book have all been either closely connected with the Tank Corps or in the Corps itself, they one and all were partakers in either the creation of the Corps or in the many actions in which it fought.

So much assistance have I received that I can at most but consider myself as editor to a mass of information provided for me by others. Those I more especially wish to thank amongst this goodly company are the following:

Captain the Hon. Evan Charteris, G.S.O.3, Tank Corps, for the accurate and careful records of the Corps which he compiled from the earliest days of the tank movement in 1914, to the close of the Battle of Cambrai. Many of these were written under, shall I say, far from luxurious circumstances, for Captain Charteris, I feel, must have often found himself, in his shell-blasted *estaminet*, less well cared for than the rats of Albert and as much out of place as Alcibiades in a Peckham parlour.

When Captain Charteris forsook the *"cabaret sans nom,"* for some ill-disposed shell had removed half the signboard, Captain O. A. Archdale, A.D.C. to General Elles, took up the difficult task and, from March 1918 onwards, kept the Tank Corps Diary upon which Chapters 29, 33, 35 and 37 are founded.

Taking now the chapters seriatim, I have to thank Major G. W. G. Allen, M.C., G.S.O.2, War Office, (previously a Tank Corps engineer officer in France), for parts of Chapter 1, and also the editors of *The American Machinist* and *The Engineer* for allowing me to quote respectively from the following admirable articles: "The Forerunner of the Tank," by H. H. Manchester, and "The Evolution of the Chain Track Tractor"; Sir Eustace Tennyson D'Eyncourt, K.C.B., Director of Naval Construction, the Admiralty, and Major-General E. D. Swinton, C.B., D.S.O., both pioneers of the tanks, and indefatigable workers in the cause, for much of the information in Chapters 2 and 4; Major H. S. Sayer, G.S.O.2, War Office, (*ibid*), for Chapter 3; Major O. A. Forsyth-Major, Second in Command of the Palestine Tank Detachment, for the reports relative to the second and third Battles of Gaza, upon which Chapters 11 and 17 are based; Major S. H. Foot, D.S.O.,

G.S.O.2, War Office, (previously Brigade Major, 2nd Tank Brigade, in France), my close friend and fearless assistant, for suggestions generally, and particularly in Chapter 16.

My thanks are also due to some unknown but far-sighted benefactor of the Tank Corps for Chapter 20; to Lieutenant-Colonel D. W. Bradley, D.S.O., and Brigadier-General E. B. Mathew-Lannowe, C.M.G., D.S.O., G.O.C. Tank Corps Training Centre, Wool, for information regarding the Depot in Chapter 21; to the relentlessly inventive Lieutenant-Colonel L. C. A. de B. Doucet, O.C. Tank Carrier Units, and so commander of the first supply fleet which ever "set sail" on land, for information to be found in Chapter 22; to Lieutenant-Colonel J. D. M. Molesworth, M.C., A.D.A.S., Tank Corps, who in spite of the scholastics gave the lie to the tag *Ex nihilo nihil fit*, for parts of Chapter 24; to Major R. Spencer, M.C., Liaison Officer, Tank Corps, whose unfailing charm and insight always succeeded in extracting from our brave Allies not only the glamour of great adventures but the detail of truthful occurrences, for the events described in Chapters 25 and 26; to Major F. E. Hotblack, D.S.O., M.C., G.S.O.2, War Office, (Previously G.S.O.2, Intelligence Headquarters, Tank Corps), my friend and companion, who unfailingly would guide *anyone* over wire and shell-hole immune and unscathed, for Chapters 28, 31, and 34; to Lieutenant C. B. Arnold, D.S.O., Commander of Whippet Tank "Musical Box," for the simple and heroic exploit related in Chapter 30; to Major T. L. Leigh Mallory, D.S.O., O.C. 8th Squadron, R.A.F., whose energy resulted not only in the cementing of a close comradeship between the two supreme mechanical weapons of the age but of a close co-operation which saved many lives in battle, for much of Chapter 32; to Lieutenant-Colonel E. J. Carter, O.C. 17th Tank Armoured Car Battalion, who was as great a terror to the German Corps Commanders as Paul Jones was to the Manchester merchantmen and who had the supreme honour to break over the Rhine the first British flag—the colours of the Tank Corps—for Chapter 38.

It was a great brotherhood, the Tank Corps, and if there were "duds" in it there certainly were not old ones, for the Commander of the Corps, Major-General H. J. Elles, C.B., D.S.O., was under forty, and most of his staff and subordinate commanders were younger than himself. Youth is apt, rightly, to be enthusiastic, and General Elles must frequently have had a trying time in regulating this enthusiasm, canalising it forward against the enemy and backward diplomatically towards our friends.

We of the Tank Corps Headquarters Staff knew what we wanted. Realising the power of the machine which the brains of England had created, we never hesitated over a "No" when we knew that hundreds if not thousands of lives depended on a "Yes."

Modestly, looking back on the war from a comfortable armchair in London, I see clearly, quite clearly, that we were right. The war has proved it, and our endeavours were not in vain. We were right, and youth generally is right, for it possesses mental elasticity, its brains are plastic and not polarised. The mental athlete is the young man: the Great War, like all other wars, has proved this again and again. We have heard much of Hindenburg and Ludendorff, but they scoffed at the tank just as Wurmser and Alvinzi scoffed at the ragged *Voltigeurs* of the Army of Italy with which the Little Corporal was, in 1796, about to astonish Europe. We have also astonished Europe, we who wandered over the Somme battlefield with dimmed eyes, and over the Flanders swamps with a lump in our throats.

There was Colonel F. Searle, C.B.E., D.S.O., Chief Engineer of the Corps, a true civilian with a well-cut khaki jacket and lion-tamer's boots. He could not understand the military ritual, and we soldiers seemed never to be able to explain it to him. Throughout the war, in spite of his immense mechanical labours, I verily believe he had only one wish, and this was to erect a guillotine outside a certain holy place. There was Major G. A. Green, M.C., Colonel Searle's deputy, the father of terrible propositions, the visitor of battlefields, the searcher after shell-holes, the breather of profane words. The Corps owed a lot to Green; a firm believer in seeing things before criticising them, he was a very great asset.

The "King of Grocers," this was Colonel T.J. Uzielli, D.S.O., M.C., D.A. and Q.M.G. of the Corps, business-like, and an administrator from boot to crown. Suave yet fearless, tactful yet truthful, the Corps owed much to his ability. It was never left in want, his decision gave it what it asked for, his prevision cut down this asking to a minimum. Ably seconded by Major H. C. Atkin-Berry, D.S.O., M.C., and Major R. W. Dundas, M.C., the "A" and "Q" branches of the Tank Corps Staff formed the foundation of the Corps' efficiency.

On the "G" side there was myself. Under me came Major G. le Q. Martel, D.S.O., M.C., very much R.E. and still more tanks, the man who "sloshed" friend or foe. One day, in March 1918, I was at Fricourt, then none too healthy. Martel walked down the road: "Where are you going?" I shouted.

"To Montauban," he answered.

"I hear it is full of Boche," I replied.

"Well, I will go and see," said Martel, and off he moved eastwards.

There was Major F. E. Hotblack, D.S.O., M.C., lover of beauty and battles, a mixture of Abelard and Marshal Ney. Were Ninon de l'Enclos alive he would have been at her elbow; as she is dust, he, instead, collected "*troddels*" (German bayonet tassels), off dead Germans—a somewhat remarkable character. As G.S.O.2 Training, Major H. Boyd-Rochfort, D.S.O., M.C., from West Meath, his enthusiasm for tanks nearly wrecked a famous corps; yet Boyd only smiled, and his smile somehow always reminded one of Peter Kelly's whisky, there was a handshake or a fight in it. The two G.S.O.s3 were Captain the Hon. E. Charteris and Captain I. M. Stewart, M.C. Charteris was the "*Arbiter Elegantiarum*" of our Headquarters. He kept the corps' records, as already stated, and without these it would scarcely have been possible to write this history. He was our *maître d'hôtel*; he gave us beach nut bacon and honey for breakfast, kept his weather eye open for a one-armed man, elaborated menus which rivalled those of Trimalchio, and gave sparkle to us all by the ripple of his wit.

Lastly, Ian Stewart of the Argyll and Sutherland Highlanders. In kilts, no girl between Hekla and Erebus has ever been known to resist him; but his efforts, whilst in the Tank Corps, did not lie in conquering hearts but in perpetually worrying my unfortunate self to become party to his own suicide—for nothing would keep him from the battlefield.

The first three brigadiers of the corps were all remarkable men. Brigadier-General C. D'A. B. S. Baker-Carr, C.M.G., D.S.O., commanding the 1st Tank Brigade, started the war as a gentleman chauffeur, a most cheery companion, the Murat of the Corps, ever ready for a battle or a game. I remember him at Montenescourt, during the Battle of Arras 1917, fighting with the telephone, at Ypres fighting with the mud, at Cambrai fighting with a comfortable, vacant, rotund little man, but ever cheerful and prepared to meet you with a smile and a glass of old brandy. Commanding the 2nd Tank Brigade was Brigadier-General A. Courage, D.S.O., M.C. He possessed only half a jaw, having lost the rest at Ypres; yet at conferences he was a host in himself, and what a "*pow-wow*" must have been like before the Boche bullet hit him is not even to be found in the works of the great Munchausen. (*From Chauffeur to Brigadier*-by C. D. Baker-Carr is also published by Leonaur.) No detail escaped his eye, no trouble was too great, and no fatigue sufficient to suggest a pause. The successes of Hamel and Moreuil in 1918 were due to his energy, and on these successes was the Battle of Amiens founded. The last of the original brigadiers was Brigadier-General J. Hardress-Lloyd, D.S.O., commanding the 3rd Tank Brigade. He started the war as a stowaway. This resulted in no one ever discovering what his substantive rank was; by degrees a myth as to his origin was cultivated by innumerable "A" clerks both in France and England; these lived and throve on this mystery, which no doubt will at a distant date be elucidated by some future Lemprière. Hardress-Lloyd was one of the main causes of the Battle of Cambrai. He, I believe, introduced the idea to General Sir Julian Byng, this away back in August 1917. Hardress-Lloyd was a man of big ideas and always kept a good table and a fine stable—in fact, a *beau sabreur*. I will leave Hardress at that.

Above are to be sought the real foundations of the corps' efficiency under its gallant Commander, Major-General H. J. Elles, C.B., D.S.O., who endowed it with that high moral, that fine *esprit de corps* and jaunty *esprit de cocarde* which impelled it from one success to another. These foundations no future historian is likely to be so intimately acquainted with as I—and now for the story. (Certain chapters of this history originally appeared in a privately circulated series of papers entitled *Weekly Tank Notes*.)

The history itself is purposely uncritical, because any criticism which might have been included is so similar to that directed against the introducers of the locomotive and the motorcar that it would be but a repetition, tedious enough to the reader, were it here repeated.

Human opinion is conservative by instinct, and what to mankind is most heterodox is that which is most novel: this is a truism in war as it is in politics or religion. It took 1000 years for gunpowder to transform war. In 1590, a certain Sir John Smythe wrote a learned work: "Certain discourses concerning the forms and effects of divers sorts of weapons, and other very important matters militarie, greatlie mistaken by divers of our men of warre in these daies; and chiefly, of the Mosquet, the Caliver, and the Long-bow; as also of the great sufficiencie, excellencie and wonderful effects of Archers," in which he extols an obsolete weapon and decries a more modern one—the arquebus. "For the reactionaries of his time George Stephenson with his locomotive was the original villan of the piece; he was received with unbridled abuse and persecution. Most of Stephenson's time was spent in fighting fools." (*How to make Railways Pay for the War*, Roy Horniman.)

At the beginning of the present century nearly every English country gentleman swore that nothing would ever induce him to exchange his carriage for a motorcar—yet the locomotive and the motorcar have triumphed, and triumphed so completely that all that their inventors claimed for them appears today as hostile criticism against their accomplishments.

So with the tank, it has come not only to stay but to revolutionise, and I for one, enthusiastic as I am, do not for a minute doubt that my wildest dreams about its future will not only be realised but surpassed, and that from its clumsy endeavours in the Great War will arise a completely new direction in the art of warfare itself.

That the Tank had, and still has, many doubters, many open critics, is true enough; but there is no disparagement in this, rather is it a compliment, for the masses of mankind are myopic, and had they accepted it with acclaim how difficult would it have been for it even to come, let alone stay and grow.

The criticism directed against this greatest military invention of the war was the stone upon which its progress was whetted. Without criticism we might still have Big Willie, but we enthusiasts determined that not only would we break down this criticism by means of the machine itself, but that we would render our very machine ridiculous by machines of a better type, and it is ridicule which kills. So, we proceeded, and as type followed type, victory followed victory. Then our critics tacked and veered: it was not the tank they objected to but our opinions regarding it; they were overstatements; why, we should soon be claiming for it powers to boil their morning tea and shave them whilst still in bed. Why not? If such acts are required, a tank can be built to accomplish them, because the tank possesses power and energy, and energy is the motive force of all things.

It is just this point that the critics missed; their minds being controlled by the conventions of the day. They could not see that if the horse-power in a man is x, that the circumference of his activities is a circle with x as its radius. They could not see that if the horse-power of a machine is $100x$; its circumference will be vastly greater than that of man's; neither could they see that whilst in man x is constant, provided the man is supplied regularly with beef, bread and beer, in a machine x may be increased almost indefinitely, and that if a circle with n as its circumference will not embrace the problem, probably all that is necessary is to add more x's to its radius.

Indeed, the science of mechanics is simplicity itself when compared with that of psychology, and as in war mechanics grow so will psychology, in comparison, dwindle, until perhaps we may see in armies as complete a change from hand-weapons to machine-weapons as we have seen in our workshops from hand-tools to machine-tools, and the economy will be as proportionate.

Before the Great War I was a believer in conscription and in the Nation in Arms; I was an 1870 soldier. My sojourn in the Tank Corps has dissipated these ideas. Today I am a believer in war mechanics, that is, in a mechanical army which requires few men and powerful machines. Equally am I a disbeliever in what a venerable acquaintance, old in ideas rather than years, said to me on the afternoon of November 11th, 1918. These are his words, and I repeat them as he exclaimed them:

"Thank God we can now get back to real soldiering!"

J. F. C. F.

Langham Hotel, London, W.1.
November 20, 1919.

Chapter 1

The Origins of the Tank

In war the main problem to solve is—"How to give blows without receiving them"; it has always been so and is likely always to remain so, for battles are two-act tragedies: the first act consisting in hitting and the second in securing oneself against being hit.

If we look back on the 4,000 years of the known history of war, we shall find that its problems are always the same: thus, in battle the soldier has to think of four main acts:

(1) How to strike his opponent when at a distance from him;

(2) How to move forward towards him;

(3) How to strike him at close quarters;

(4) How to prevent himself being struck throughout the whole of this engagement.

In these four acts must be sought the origins of the tank, the idea of which is, therefore, much older than the Trojan horse; indeed, it dates back to some unknown period when aboriginal man raised his arm to ward off the blow of an infuriated beast or neighbour.

To ward off a blow with the bare skin is sometimes a painful operation; why not then cover the arm with leather or iron, why not carry a shield, why not encase the whole body in steel so that both arms instead of one may be used to hit with, for then man's offensive power will be doubled?

If we look back on the Middle Ages, we find that such a condition of fighting was actually possible and that knights clad in armour *cap-à-pie* were practically invulnerable. As regards these times there is an authentic record concerning twenty-five knights in armour who rode out one day and met a great mob of insurgent peasants which they charged and routed, killing and wounding no fewer than 1,200 of them, without sustaining a single casualty themselves. To all intents and purposes, these knights were living tanks—a combination of muscular energy, protective armour, and offensive weapons.

Knights in armour remained practically invulnerable as long as the propellant for missile weapons was limited to the bow-string and as long as the knights fought within the limitations which their armour imposed upon them. At Crécy and similar battles, the chivalry of France suffered defeat more through the condition of ground they attempted to negotiate, than through the arrows of the English archers. They, in fact, became "ditched" like a tank in the mud, and being rendered immobile, fell an easy prey to the enemy's men-at-arms. A fact which proves that it was not the arrow which generally destroyed the knight is that the archers were equipped with maces or leaden hammers by means of which the knight could, when once bogged or "bellied," be stunned, rendered innocuous, his armour opened, and he himself taken prisoner for ransom.

******

The arrow was the means of immobilising the knight by forcing him to dismount. Horse armour was never very satisfactory. Regarding the maces, a chronicler writes of their use by the archers at Agincourt: "It seemed as though they were hammering upon anvils."

******

The true banisher of armour was gunpowder, for when once the thickest armour, which human energy would permit of being worn, could be penetrated, it became but an encumbrance to its wearer. Though gunpowder was introduced as a missile propellant on the battlefield as early as the twelfth century, it was not until the close of the fourteenth and beginning of the fifteenth centuries that its influence began to be felt, and it is interesting to note that directly it became apparent that the hand gun would beat armour carried by men, other means of carrying it were introduced. These means took the form of battle cars or mobile fortresses.

******

The idea of a mobile fortress or battle car is very old: chariots are known to have existed in Assyria as far back as the year 3500 B.C. The Egyptians and Israelites both adopted them from this source. In Biblical times their tactical utility was considerable, as the Book of Judges relates. The Chinese, as early as 1200 B.C., made use of war cars armoured against projectiles.

******

Conrad Kyeser, in his military manuscript, written between 1395

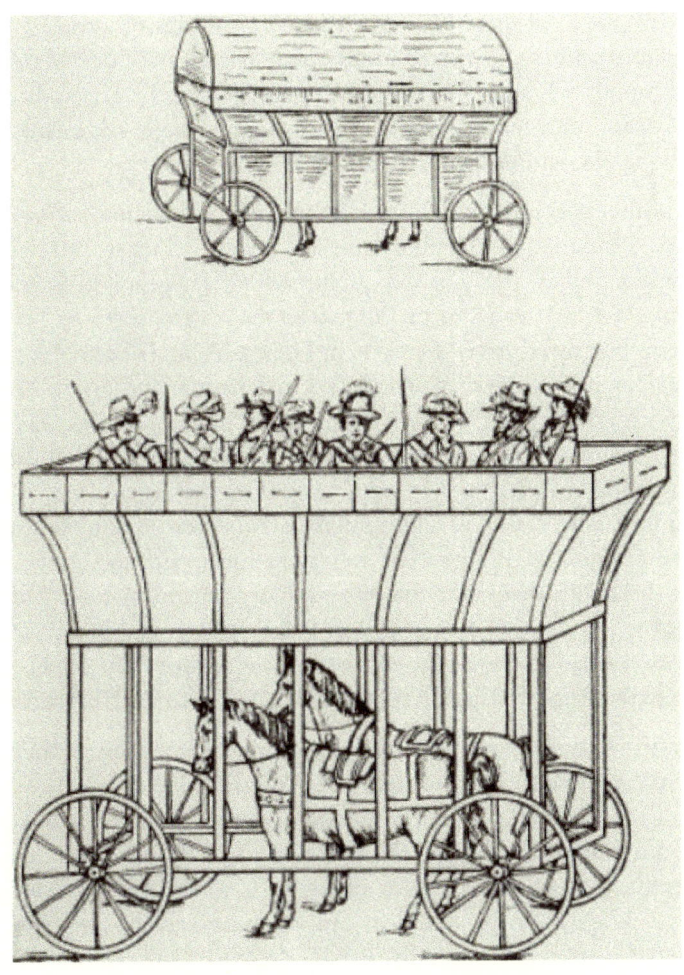

Diagram 1. Scottish War Cart, 1456.

and 1405, pictures several "battle cars." (Much of the following information is taken from an article entitled "The Forerunner of the Tank," by H. H. Manchester, published in *The American Mechanist*, vol. 49, No. 15.) Some of these are equipped with lances, whilst others are armed with cannon. A few years later, in 1420, Fontana designed a large "battle car," and the following year Archinger another, to enclose no fewer than 100 men. All these cars were moved by means of muscle power, *i.e.* men or animals harnessed inside them. A picture of one of these is to be found in Francis Grose's *Military Antiquities*, vol. I, (see diagram 1). Its crew consisted of eight men, the same as the Mark I Tank. The following extract concerning these carts is of interest:

> Another species of artillery were the war carts, each carrying two Peteraros or chamber'd pieces; several of these carts are represented in the Cowdry picture of the siege of Bullogne, one of which is given in this work; these carts seem to have been borrowed from the Scotch; Henry, in his *History of England*, mentions them as peculiar to that nation, and quotes the two following acts of parliament respecting them; one *A.D.* 1456 wherein they are thus described: 'it is tocht speidfull that the King mak requiest to certain of the great burrows of the land that are of ony myght, to mak carts of weir, and in elk cart twa gunnis and ilk one to have twa chalmers, with the remnant of the graith that effeirs thereto, and an cunnard man to shute thame.' By another Act, *A.D.* 1471, the prelates and barons are commanded to provide such carts of war against their old enemies the English (Black Acts, James II, Act 52, James III, Act 55).

With all these war carts the limitations imposed upon them by muscular motive force must have been considerable on any save perfectly firm and level ground, consequently other means of movement were attempted, and during the last quarter of the fifteenth century the battle car enters its second phase. In a work of Valturio's dated 1472, a design is to be found of one of these vehicles propelled by means of wind wheels (see diagram 2). Ten years later we find Leonardo da Vinci engaged in the design of another type of self-moving machine. Writing to Ludovico Sforza he says:

> I am building secure and covered chariots which are invulnerable, and when they advance with their guns into the midst of the foe, even the largest enemy masses must retreat; and behind

them the infantry can follow in safety and without opposition.

What the motive force of this engine of war was is unknown, but the above description is that of the tank of today, in fact so accurate is this description that Leonardo da Vinci, nearly 350 years ago, had a clearer idea of a tank operation than many a British soldier had prior to the Battle of Cambrai, fourteen months after the first tank had taken the field.

A somewhat similar self-moving wagon was designed for Maximilian I and in 1558 Holzschuher describes a battle car a picture of which shows it in action preceded by infantry and flanked by cavalry (see diagram 3).

In 1599 Simon Stevin is supposed to have constructed for the Prince of Orange two veritable landships; these consisted in small battleships fully rigged, mounted upon wheels (see diagram 4).

> The earliest English patent for a self-moving wagon which could, if desired, be used in war, was probably that taken out by David Ramsey in 1634. In 1658 Caspar Schott designed one to enclose 100 men and to be employed against the Turks. (*The Forerunner of the Tank*, by H. H. Manchester.)

All the users of these inventions were destined to disappointment, for the science of mechanics was not sufficiently advanced to render self-movement practical and it was not until the middle of the eighteenth century that a fresh attempt was made to reintroduce so essential a weapon as the war cart. The following account of this reintroduction is quoted from Mr. Manchester's most interesting article:

> After the practical application of steam by Watt in 1765 we find an early attempt to apply it to land transportation in what must be considered the first steam automobile. As early as 1769 Cugnot in France set a steam boiler upon the frame of a wagon and succeeded in making the wagon go. His idea was that this invention could be used in war, and on this presumption, he was the next year assisted by the government to construct an improvement. The speed, however, was scarcely more than 2½ miles an hour, and the machine would run only twenty minutes before it had to stop for fifteen minutes to get up more steam. In his first public trial he had the ill-luck to run into and knock down part of a stone wall. This led to his being temporarily cast into jail, and his experiments were abandoned. Napoleon

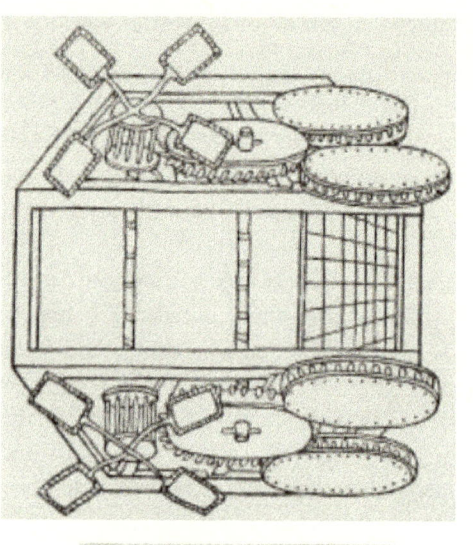

Diagram 2. Valturio's War Chariot, 1472.

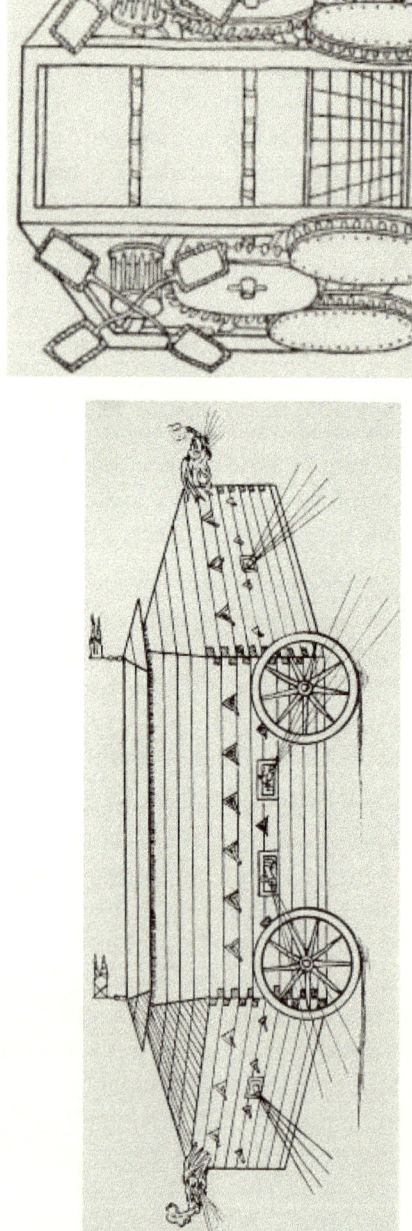

Diagram 3. Holzschuher's Battle Car, 1558.

must have visualised the possibilities of Cugnot's machine for military purposes, for when the great general was selected a member of the French Institute, the subject of his paper was 'The Automobile in War.'

The "battle car" had now, at least experimentally, evolved into the steam wagon which could run on roads; the next step was to invent one which would move in any direction across country, in other words to replace the wheels by tracks. The evolution of the caterpillar tractor brings us to the fourth phase in the evolution of the "battle car."

The idea of distributing the weight of a vehicle over a greater area than that provided by its own wheels is by no means a novel one; one year after Cugnot produced the first steam automobile Richard Lovell Edgeworth patented a device whereby a portable railway could be attached to a wheeled carriage; it consisted of several pieces of wood which moved in regular succession in such a manner that a sufficient length of railing was constantly at rest for the wheels to roll upon. The principle of this device was but a modification of that upon which the tracks of tanks now depend, and all subsequent ideas were founded on this basis. (For Edgeworth's invention and the short summary of the footed-wheel, etc., which follows see *The Engineer*, August 10, 1917, and following issues.)

The endless chain track passed through various early patents. In 1801 Thomas German produced "a means of facilitating the transit of carriages by substituting endless chains or a series of rollers for the ordinary wheels." This definitely cut adrift from the idea of wheels and replaced it by that of tracks. In 1812 William Palmer produced a somewhat similar invention, and in 1821 John Richard Barry patented a contrivance consisting of two endless pitched chains, stretched out and passing round two chain wheels at the end of the carriage, one on each side, which formed the rails or bearing surface of the vehicle.

Footed wheels were not, however, abandoned, and in 1846 a picture of the Boydell engine shows the wheels of this machine fitted with feet. In 1861 an improved wheel-foot was patented by Andrew Dunlop which was modified by other inventors and by degrees evolved into the pedrail, trials of which were carried out at Aldershot under the War Office in 1905.

In 1882 Guillaume Fender of Buenos Aires suggested and John Newburn patented certain improvements to endless tracks. Fender realised that the attempts to produce endless travelling railways had

Diagram 4. Simon Stevin's Landship, 1599.

Diagram 5. The Applegarth Tractor, 1886.

not met with great success owing to the shortness of the rails or tracks employed; he, therefore, proposed that their length should be the same as the distance between the vehicle's axles. If it were desired to have short links the number of wheels must be increased; furthermore, should the tractor be used for hauling a train of wagons, the endless track should be long enough to embrace all the wheels. This is the original idea of the all-round track.

Among the many interesting patents of about this date were the Applegarth tractor of 1886 (see diagram 5) and the Batter tractor of 1888. In the former the forward portion of the track was inclined and suggests the contour of the track as applied to the front of tanks. The track being raised in front gives an initial elevation when an obstacle is met with and very greatly assists in surmounting banks and other irregularities.

Diagram 6 depicts the Batter tractor and it clearly shows the basic ideas which have been employed in tank transmission and tank design. This tractor was patented in the U.S.A., it was furnished with two tracks, their contour very closely resembling those of the Medium Mark "A" (Whippet) and gun-carrier machines (see Plates 3 and 7) The motive power was steam, and two separate engines, fed by one boiler, were used, one to drive each track; apparently provision was made, if desired, for the crankshafts of these engines to be clutched together. Each track consisted of two endless belts, an inner and an outer; the outer belt, that which impinges on the ground, was composed of shoes arranged transversely and coupled together. Between the outer belt and the rollers ran the inner belt. The inner belt or link was of much less width than the outer and thus allowed the latter to swivel and adapt itself to irregularities of the ground, whilst the working of the rollers was not interfered with. A system almost identical with this one has recently been adopted for tank tracks.

The rollers were alternately flanged and plain, as on tanks. Two tails for steering and balancing the machine were fitted; a similar idea was adopted on Mark I machines and gun carriers, but subsequently discarded.

The general introduction of the internal-combustion engine and petrol as a fuel gave a further impetus to the tracked machine. In 1900 Frank Bramond patented a track which could be applied to pneumatic-tyred vehicles, either to single wheels or to two pairs of wheels. In 1907 a Rochet-Schneider was fitted with a track by Roberts and tested at Aldershot. This car was exhibited together with a 70

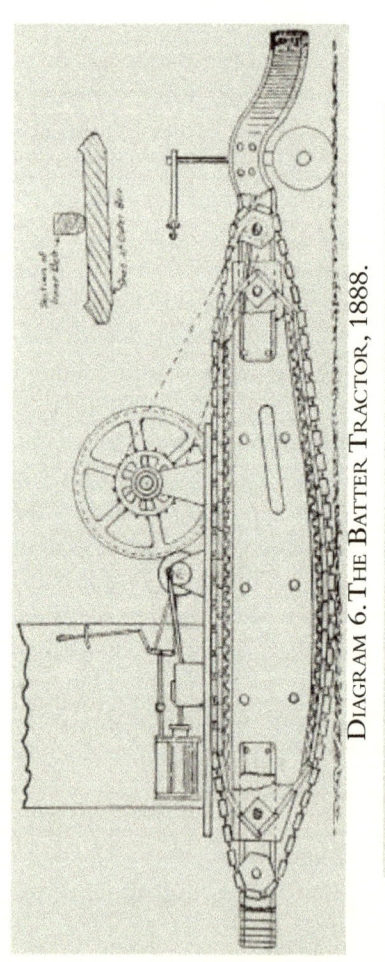

Diagram 6. The Batter Tractor, 1888.

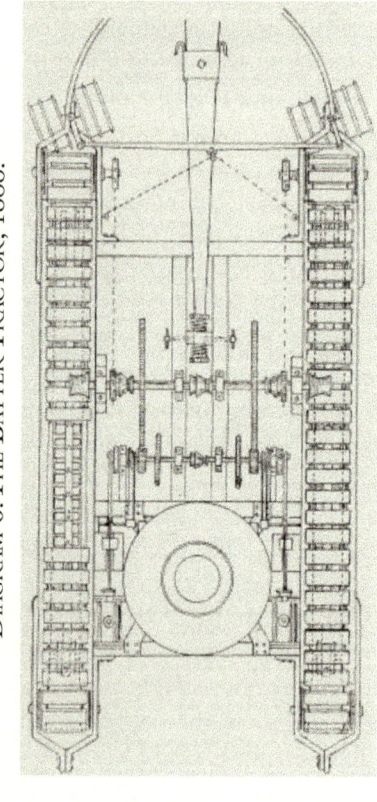

Diagram 6a. The Batter Tractor, 1888.

h.p. Hornsby chain-track tractor and took part in the Royal Review at Aldershot in May 1908. This same year Hornsby fitted up a 75 h.p. Mercedes motorcar with a track to demonstrate its advantages for high-speed work on sand.

> This car was run daily for five months at Skegness, on loose sand, and it is understood that a speed of twenty miles an hour was obtained. (*The Engineer, ibid.*)

Of later years, American inventors and manufacturers have made great progress in chain-track tractors, but practically all the principles of design were originally applied in Great Britain. The Holt caterpillar is the outstanding American design for tractors which has been adopted during the war.

It is interesting to note with reference to the above inventions that neither Germany nor Austria ever appears to have contributed any basic suggestion relating to track-driven machines.

To return now to the military aspect of our subject, gunpowder did away with armour, for if armour can be pierced its defensive value is lost and it only becomes an encumbrance to the wearer by reducing his mobility and exhausting his muscular energy. Did this change the main problem in the art of war? Not at all, for "the giving of blows without receiving them" remains the unchangeable object of battle irrespective of the change of weapons, and all that happened was, that the soldier, no longer being able to seek protection by body-armour, sought it elsewhere—by manoeuvring, by covering fire and entrenchments as typified in the drill of Frederick the Great, the cannonades and sharpshooters' fire of Napoleon, the fortifications of Vauban, and later on the use of ground by Wellington as cover from fire.

The opening of the war in 1914 saw all sides equipped with similar weapons and in comparatively similar proportions. The great sweep of the Germans through Belgium was followed by the Battle of the Marne, a generic term for a series of bloody engagements which raged from Lorraine to Paris. Then came the great reaction—the German retreat to the Aisne, the heights along which had been hastily prepared for defence. The battle swayed whilst vigour lasted and then stabilised as exhaustion intervened. At first cautiously, then rapidly, did the right flank of the German Armies and the left flank of those of the French and British seek to out-manoeuvre each other. This led to the race for the coast. Meanwhile came the landing of the British 7th Division at Zeebrugge and then the First Battle of Ypres, which closed the German offensive on the British front for three years and four months.

The quick-firing field-gun and the machine-gun, used defensively, proved too strong for the endurance of the attackers, who were forced to seek safety by means of their spades, rather than through their rifles. Whole fronts were entrenched, and before the end of 1914, except for a few small breaks, a man could have walked by trench, had he wished to, from Nieuport almost into Switzerland.

With the trench came wire entanglements—the horror of the attack, and the trinity of trench, machine-gun, and wire, made the defence so strong that each offensive operation in turn was brought to a standstill.

The problem which then confronted us was a twofold one:

Firstly, how could the soldier in the attack be protected against shrapnel, shell-splinters, and bullets? Helmets were reintroduced, armour was tried, shields were invented, but all to no great purpose.

Secondly, even if bullet-proof armour could be invented, which it certainly could, how were men laden down with it going to get through the wire entanglements which protected every position?

Three definite solutions were attempted—the first, artillery; the second, gas; and the third, tanks—each of which is a definite answer to our problem if the conditions are favourable for its use. Thus, at the Battle of the Dunajec, in the spring of 1915, the fire of Mackensen's massed artillery smashed the Russian front; this success being due as much to the fewness of the Russian guns as to the skill of that great soldier. At the Second Battle of Ypres the German surprise gas attack succeeded because the British and French possessed no antidote. At the First Battle of Cambrai, the use of tanks on good firm ground proved an overwhelming success, whilst at the Third Battle of Ypres, on account of the mud, they were an all but complete failure.

All armies attempted the first method by increasing the number of their guns, the size of their guns, and the quantity of their ammunition. So thoroughly was this done that whole sectors of an enemy's front were blasted out of recognition. This, however, was only accomplished after all surprise had been sacrificed by obvious preparation during which notice and time were given to the enemy to mass his reserves in order to meet the attack. Further than this, though the enemy's wire and trenches were destroyed all communications on his side of "No Man's Land" were obliterated, with the result that a new obstacle, "the crumped area," proved as formidable an antagonist to a continuous advance, by hampering supply, as uncut wire had done to a successful assault, by forbidding infantry movement.

Instead of solving the problem: "How could mobility be reintroduced on the Western Front?" the great increase in artillery, during 1915 and 1916, only complicated it, for, though the preliminary bombardment cut the wire and blew in the enemy's trenches and the creeping barrage protected the infantry in a high degree, every artillery attack during two years ended in failure due to want of surprise at its initiation and the impossibility of adequate supply during its progress.

The Germans attempted the second method—gas, and from the Second Battle of Ypres the chemist fell in alongside the soldier. That gas might have won the war is today too obvious to need accentuation. Two conditions were alone requisite—sufficient gas and a favourable wind. Fortunately for us the German did not wait long enough to manufacture gas in quantity; unfortunately for them the prevailing wind on the Western Front is westerly, consequently when we and the French retaliated, they got more than they ever gave us.

The introduction of gas still further complicated the problem, for, whilst it is easy for the defender to launch gas clouds, it is difficult for an attacker to do so, consequently once soldiers had been equipped with respirators the defence gained by this method of fighting and warfare became still more immobile.

As regards the British front the opening day of the First Battle of the Somme, July 1, 1916, showed, through the terrible casualty lists which followed, how far the defence had become the stronger form of war. At no date in the whole history of the war was a stalemate termination to all our endeavours more certain. The hopes of nearly two years were shattered in a few hours before the ruins of Thiepval, Serre, and Gommecourt, where our men fell in thousands before the deadly machine-gun fire of the enemy. Eleven weeks later, on September 15, a solution to the problem became apparent, a solution due to the efforts of a small band of men, of whose energy and endeavours the next chapter will relate.

CHAPTER 2

The Invention of the Landship

It is not proposed in this chapter to give an answer to the question: "Who first thought of the tank?" The idea of combining mobility with offensive power and armour, as the previous chapter has shown, is a very old one, so old and so universal throughout history that, when the Great War broke out in 1914, many soldiers and civilians alike must have considered ways and means of reintroducing the knight in armour and the battle car by replacing muscular energy by mechanical force—in other words, by applying petrol to the needs of the battlefield.

During August and September 1914, armoured cars had been employed with considerable success in Belgium and north-western France. This no doubt brought with it the revival of the idea. Be this as it may, in October of this year Lieutenant-Colonel (now Major-General) E. D. Swinton put forward a suggestion for the construction of an armoured car on the Holt tractor or a similar caterpillar system, capable of crushing down wire entanglements and crossing trenches.

At the same time, Captain T. G. Tulloch, manager of the Chilworth Powder Company, was also devoting his attention to the possibility of constructing a land cruiser sufficiently armoured to enable it to penetrate right up to the enemy's gun and howitzer positions. In November the idea was communicated by Captain Tulloch to Lieutenant-Colonel Swinton and to Lieutenant-Colonel (now Colonel Sir Maurice) Hankey, Secretary to the "Committee of Imperial Defence," and later on to Mr. Churchill, then First Lord of the Admiralty, who, in January 1915, wrote his now historic letter to Mr. Asquith:

> My Dear Prime Minister,
> I entirely agree with Colonel Hankey's remarks on the subject of special mechanical devices for taking trenches. It is extraordinary that the Army in the field and the War Office should have allowed nearly three months of warfare to progress without addressing their minds to its special problems.
> The present war has revolutionised all military theories about the field of fire. The power of the rifle is so great that 100 yards is held sufficient to stop any rush, and in order to avoid the severity of the artillery fire, trenches are often dug on the reverse slope of positions, or a short distance in the rear of villages, woods, or other obstacles. The consequence is that the war has become a short-range instead of a long-range war as was expected, and opposing trenches get ever closer together, for mutual safety from each other's artillery fire.
> The question to be solved is not, therefore, the long attack over a carefully prepared glacis of former times, but the actual getting across 100 or 200 yards of open space and wire entangle-

ments. All this was apparent more than two months ago, but no steps have been taken and no preparations made.

It would be quite easy in a short time to fit up a number of steam tractors with small armoured shelters, in which men and machine-guns could be placed, which would be bullet-proof. Used at night they would not be affected by artillery fire to any extent. The caterpillar system would enable trenches to be crossed quite easily, and the weight of the machine would destroy all wire entanglements.

Forty or fifty of these engines, prepared secretly and brought into positions at nightfall, could advance quite certainly into the enemy's trenches, smashing away all the obstructions and sweeping the trenches with their machine-gun fire, and with grenades thrown out of the top. They would then make so many *points d'appui* for the British supporting infantry to rush forward and rally on them. They can then move forward to attack the second line of trenches.

The cost would be small. If the experiment did not answer, what harm would be done? An obvious measure of prudence would have been to have started something like this two months ago. It should certainly be done now.

The shield is another obvious experiment which should have been made on a considerable scale. What does it matter which is the best pattern? A large number should have been made of various patterns; some to carry, some to wear, some to wheel. If the mud now prevents the working of shields or traction engines, the first frost would render them fully effective. With a view to this I ordered a month ago twenty shields on wheels, to be made on the best design the Naval Air Service could devise. These will be ready shortly, and can, if need be, be used for experimental purposes.

A third device, which should be used systematically and on a large scale, is smoke artificially produced. It is possible to make small smoke barrels which, on being lighted, generate a great volume of dense black smoke, which could be turned off or on at will. There are other matters closely connected with this to which I have already drawn your attention, but which are of so secret a character, that I do not put them down on paper.

One of the most serious dangers that we are exposed to is the possibility that the Germans are acting and preparing all these surprises, and that we may at any time find ourselves exposed to some entirely new form of attack. A committee of engineering officers and other experts ought to be sitting continually at the War Office to formulate schemes and examine suggestions, and I would repeat that it is not possible in most cases to have lengthy experiments beforehand.

If the devices are to be ready by the time, they are required it is indispensable that manufacture should proceed simultaneously with experiments. The worst that can happen is that a comparatively small sum of money is wasted.

Yours, etc.

At about the time that the above letter was written, Lieutenant-Colonel Swinton again brought the matter forward and urged the desirability of action being taken, but as it was stated that the design and building of Captain Tulloch's machine would take a year to complete it appears that this led to the proposals being shelved for the time being.

On June 1, 1915, Lieutenant-Colonel Swinton, who had then returned to France, submitted an official memorandum on the above subject to G.H.Q., which was passed to Major-General G. H. Fowke, Engineer-in-Chief, for his expert opinion. This memorandum may be summarised as follows:

The main German offensive was taking place in Russia; consequently, in order to attain a maximum strength in the east, it was incumbent on the Germans to maintain a minimum one in the west; and, in order to meet the shortage of men on the Western Front, the Germans were mainly basing their defence on the machine-gun.

The problem, consequently, was one of how to overcome the German machine-gunners. There were two solutions to this problem:

(1) Sufficient artillery to blast a way through the enemy's lines.

(2) The introduction of armoured machine-gun destroyers.

As regards the second solution Lieutenant-Colonel Swinton laid down the following requirements: Speed, 4 miles per hour; climbing power, 5 ft.; spanning power, 5 ft.; radius of action, 20 miles; weight, about 8 tons; crew, 10 men; armament, 2 machine-guns and one light Q.F. gun. Further, he suggested that these machines should be used in a surprise assault having first been concealed behind our own front line in specially constructed pits about 100 yards apart. In this paper it was also pointed out that these destroyers would be of great value in a gas attack, as they would enable the most scientific means of overcoming gas to be carried.

The above memorandum was favourably considered by Sir John French, then Commander-in-Chief in France, and, on June 22, was submitted by him to the War Office with a suggestion that Lieutenant-Colonel Swinton should visit England and explain his scheme more fully.

While Lieutenant-Colonel Swinton and Captain Tulloch were urging their proposals, a third scheme was brought forward by Admiral Sir Reginald Bacon in connection with which the Secretary of State, in January 1915, ordered trials to be carried out with a 105 h.p. Foster-Daimler tractor fitted with a bridging apparatus for crossing trenches. At about the same time similar trials were made with a 120 h.p. Holt caterpillar tractor at Shoeburyness in connection with Captain Tulloch's scheme. Both experiments proved a failure.

The position, therefore, in June, so far as the army was concerned, was as follows: Proposals had been put forward by Lieutenant-Colonel Swinton, Admiral Bacon, and Captain Tulloch, and submitted to the War Office. Certain trials had been made, the result of which, in the view of the authorities, was to emphasise the engineering and other difficulties to be overcome. It was only in June 1915 that Major-General Sir George Scott-Moncrieff, Director of Fortifications and Works, War Office, who, throughout the initial period, had shown a strong interest in the development of the idea, ascertained that investigations on similar lines were being carried out by the Admiralty; he at once proposed that a "Joint Naval and Military Committee" should be formed for the purpose of dealing with the subject generally. This Committee was constituted on June 15.

The work done by the Admiralty had so far been independent. In February 1915, Mr. Churchill sent to Mr. E. H. T. (now Sir Eustace) D'Eyncourt, Director of Naval Construction, a copy of the notes embodying the proposals set forth by Major T. G. Hetherington (18th Hussars), R.N.A.S., for a new type of war machine. This machine may be described as a veritable Juggernaut, heavily armoured, highly offensive, and capable of moving across country.

It consisted of a platform mounted on three wheels, two driving wheels in front and the steering wheel behind. It was to be equipped with three turrets each containing two 4-in. guns and its motive power was to be derived from a 800 h.p. Sunbeam Diesel set of engines.

The problem of design was examined by the Air Department engineers and the following rough data, worked out at the time, are of interest:

Armament	3 twin 4 in. turrets with 300 rounds per gun.
Horse power	800 h.p. with fuel for 24 hours.
Total weight	300 tons.
Armour	3 in.
Diameter of wheels	40 ft.
Tread of main wheels	13 ft. 4 in.
Tread of steering wheels	5 ft. 0 in.
Overall length	100 ft.
Overall width	80 ft.
Overall height	46 ft.
Clearance	17 ft.
Top speed on good going	8 miles per hour.
Top speed on bad going	4 miles per hour.

The cross-country qualities of the machine it was considered would prove good. It could not be bogged in any ground passable by cavalry; it could pass over water obstacles having good banks and from 20 ft. to 30 ft. width of waterway; it could ford waterways 15 ft. deep if the bottoms were good, and negotiate isolated obstacles up to 20 ft. high. Small obstacles such as banks, ditches, bridges, trenches, wire entanglements, and ordinary woodland it could roll over easily.

Mr. D'Eyncourt considered this proposal, but coming to the conclusion that the machine would weigh more than 1,000 tons, it became apparent to him that its construction was not a practical proposition.

Mr. D'Eyncourt pointed this out to Mr. Churchill and suggested that Major Hetherington's machine should be replaced by one of a smaller and less ambitious type. To this Mr. Churchill agreed, and to deal with this question a "Landships Committee" was formed consisting of the following gentlemen:

Chairman.
Mr. D'Eyncourt.
Members.
Major Hetherington, Colonel Dumble, Mr. Dale Bussell
(appointed later).
Consultant.
Colonel R. E. Crompton.
Secretary (appointed later).
Lieutenant Stern.

Prior to the formation of this Committee another proposal had been set on foot. About November 14, 1914, Mr. Diplock of the Pedrail Company had put forward certain suggestions for the use of the pedrail for the transportation of heavy guns and war material over rough ground. After interviewing Lord Kitchener, who saw no utility in the suggestion, Mr. Diplock was referred to the Admiralty and there saw Mr. Churchill, who, taking up the matter with interest, suggested that a one-ton truck should be brought to the Horse Guards Parade for his inspection. Major Hetherington undertook to arrange this, and on February 12, 1915, a demonstration of the Pedrail machine took place.

This so impressed Mr. Churchill that he decided that a pedrail armoured car should be built.

The "Landships Committee" communicated with Messrs. William Foster, Ltd., of Lincoln, who were already making heavy tractors for the Admiralty, and Mr. (now Sir William) Tritton, their manager, was asked to collaborate in evolving two designs:

The first of the wheel tractor type.

The second of the pedrail type

.....the latter being the alternative recommended by the chairman and the Pedrail Company.

Both these designs seemed to have some promising features. The First Lord, on March 26, approved of an order being placed for twelve of the pedrail type and six of the wheel type.

The design of the pedrail machine was produced by the Pedrail Company; its length was 38 ft., its width 12 ft. 6 in., and height 10 ft. 6 in. The most interesting feature connected with this machine was that it was mounted on two bogies one behind the other, steering being rendered possible by articulating these bogies in the same horizontal plane, which gave an extreme turning radius of 65ft.

After Mr. Churchill's resignation from the Admiralty the production of the twelve pedrail cars was abandoned in spite of the fact that the engines and most of the material had been provided.

The design work was, however, continued under the direction of the "Landships Committee," and, a little later on, caterpillar tractors for experimental purposes were obtained from America. In the meantime, the question of design was discussed with Mr. Tritton, and at the same time Lieutenant (now Major) W. G. Wilson, an experienced engineer, was brought in as consultant, and a design was evolved which eventually embodied the form finally adopted and adhered to for tanks. Thus, it was through the "Landships Committee," at a moment when the military authorities were inclined to regard the difficulties connected with the problem as likely to prove insuperable, that the landship or "tank," as it was later on called, was first brought into being.

After the formation of the "Joint Naval and Military Committee" on June 15, it was agreed, as the result of correspondence between the Admiralty and War Office, that the experimental work on the landship should be taken over as a definite military service in the department of the Master-General of Ordnance. It was further agreed that the Director of Fortifications and Works should be president of the Committee, that the chairman and members of the existing "Landships Committee" should continue to serve as long as their assistance was required, and that the late First Lord of the Admiralty, Mr. Churchill, should remain in touch with the design and construction of the machines during their experimental stage. The members nominated for the Committee by the War Office were Colonel Bird of the General Staff, Colonel Holden, A.D.T., and Major Wheeler of the M.G.O.'s Department.

Early in July, Mr. Lloyd George, Minister of Munitions, discussed with Mr. Balfour, now First Lord of the Admiralty, the transference of the production of the machines from the Admiralty to the Ministry of Munitions. It was, however, subsequently decided that the Admiralty should be responsible for the production of the first trial machine, the Director of Naval Construction being responsible for the completion of the machine. This was strongly urged by Sir George Scott-Moncrieff.

In July 1915, Lieutenant-Colonel Swinton returned to England to take over the duties of assistant secretary to the "Committee of Imperial Defence." He at once took in hand the co-ordination of the various private and official efforts which were being made at this time in relation to the design of caterpillar tractors. Early in September he visited Lincoln and inspected a machine known as Little Willie, and on the 10th of this month wrote to Major Guest, Secretary of the "Experiments Committee" at G.H.Q., as follows:

> The naval people are pressing on with the first sample caterpillar ... they have succeeded in making an animal that will cross 4 ft. 6 in. and turn on its own axis like a dog with a flea in its tail. ...

In spite of its agility this machine was rejected in favour of Big Willie, a model of which was being constructed under the direction

PLATE 1

LITTLE WILLIE.

MARK IV TANK (FEMALE).

of the "Joint Committee" on the lines of the machine designed by Mr. Tritton and Lieutenant Wilson and the requirements of which had been outlined by Lieutenant-Colonel Swinton in his memorandum of June 1.

As regards these requirements, on the day following the above letter the "Experiments Committee" G.H.Q. sent the following tactical suggestions, arising out of Colonel Swinton's original proposal, to the secretary of the "Committee of Imperial Defence." They are worth quoting as they embody several of the characteristics which were introduced in the Mark I tank.

(1) The object for which the caterpillar cruiser or armoured fort is required is for employment in considerable numbers in conjunction with or as an incident in a larger and general attack by infantry against an extended front.

(2) As a general principle, it is desirable to have a large number of small cruisers rather than a smaller number of large ones.

(3) The armour of the cruiser must be proof against concentrated rifle and machine-gun fire, but not proof against artillery fire. The whole cruiser should be enclosed in armour.

(4) The tactical object of the cruiser is attack, its armament should include a gun with reasonable accuracy up to 1,000 yards, and at least two Lewis guns, which can be fired from loopholes to flank and to rear.

(5) The crew to consist of six men—two for the gun, one for each Lewis gun, and two drivers.

(6) The caterpillar must be capable of crossing craters produced by the explosion of high-explosive shell, such craters being of 12 ft. diameter, 6 ft. deep, with sloping sides; of crossing an extended width of barbed-wire entanglements; and of spanning hostile trenches with perpendicular sides and of 4 ft. in breadth.

(7) The cruiser should be capable of moving at a rate of not less than 2½ miles per hour over broken ground, and should have a range of action of not less than six hours consecutive movement.

(8) The wheels of the cruiser should be on either the "Pedrail" system or the "Caterpillar" system; whichever is the most suitable for crossing marshy and slippery ground.

Most of these requirements had already been embodied in the wooden model of Big Willie, which, when completed, was inspected at Wembley on September 28. This model was accepted as a basis on which construction was to proceed, it was in fact the first "mock up" of the eventual Mark I machine.

Big Willie was about 8 ft. high, 26 ft. long, and 11 ft. wide without sponsons, and 3 ft. wider when these were added. His armament consisted in two 6-pounder guns and two machine-guns, and the crew suggested was 1 officer and 9 other ranks.

On the following day the "Joint Committee" assembled at the Admiralty and decided that the following specifications should be worked to: weight 22 tons, speed 3½ miles per hour, spanning power 8 ft., and climbing power 4½ ft.

On December 3, Mr. Churchill addressed a paper to G.H.Q., entitled "Variants of the Offensive," in which he accentuated the necessity of concentrating more than we had done on "the attack by armour," the chief purpose of armour being to preserve mobility. He suggested the combined use of the caterpillar tractor and the shield. The caterpillars were to breach the enemy's line and then turn right and left, the infantry following under cover of bullet-proof shields. It was further suggested that the attack might be carried out at night under the guidance of searchlights. The rest of this paper dealt with "Attack by Trench Mortars, Attack by the Spade, and The Attack on the First Line."

On Christmas Day 1915, Sir Douglas Haig, who had recently taken over command of the Expeditionary Force in France, read this paper, and wishing to know more about the caterpillars mentioned, Lieutenant-Colonel H. J. Elles (later on G.O.C. Tank Corps) was sent to England to ascertain the exact position. On January 8 this officer reported in writing to G.H.Q., as follows:

> There are two producers of landships:
>
> (a) Trench Warfare working alone.
> ******
> Note:—The machine constructed by the Trench Warfare Department was the double bogey car designed by the Pedrail Company, of which it will be remembered twelve were originally ordered by the "Landships Committee" and eventually abandoned. The resuscitation of this machine arose as follows: During the summer of 1915 the Trench Warfare Department approached the Pedrail Company concerning the design of a flame projector with the capacity of 12,000 gallons of petrol.

In order to carry this weapon, the Pedrail Company suggested their original design, which, though it was not approved of by the "Landships Committee," was accepted by the Trench Warfare Department. One machine was placed on order and built at Bath by Messrs. Stothert and Pitt, the pedrails being manufactured by the Metropolitan Carriage, Wagon and Finance Co., Ltd., and the frame by Messrs. William Arrol. The machine when built weighed 32 tons unloaded, was equipped with two 100 h.p. Astor engines, and when tested out on Salisbury Plain attained a speed of 15 miles an hour. Only one of these machines was made, as eventually the idea of using mechanically driven flame projectors was abandoned.

✶✶✶✶✶✶

(b) The Admiralty Landship Committee working with the War Office.

The first have not yet made a machine, but its proposed size is 10 ft. high, 14 ft. 6 in. wide, and 36 ft. long; the second was in process of being made" (*i.e.* Big Willie).

Up to December 20, 1915, the whole cost of the experimental work had been defrayed by the Admiralty, which had also provided the personnel in the shape of No. 20 Squadron, R.N.A.S., for carrying out the work. The Admiralty had in fact fathered and been responsible for the landship since its first inception.

On December 24 the following recommendations were formulated at a Conference held at the offices of the "Committee of Imperial Defence":

Supply of Machines

(1) That if and when the Army Council, after inspection of the final experimental land cruiser, decide that such machines shall be entrusted to a small 'Executive Supply Committee,' which, for secrecy, shall be called the 'Tank Supply Committee,' and shall come into existence as soon as the decision of the Army Council is made.

✶✶✶✶✶✶

Note:—This is the first appearance of the word "tank" in the history of the machine. Up to December 1915, the machines now known as "tanks" were, in the experimental stage, called "landships" or "land cruisers," and also "caterpillar machine-gun destroyers." On December 24, whilst drafting the above report of the Conference it occurred to Colonel Swinton that the use of the above names would give away a secret which it was important to preserve. After consultation with Lieutenant-Colonel W. Daily-Jones, assistant secretary of the "Committee of Imperial Defence," the following names were suggested by Colonel Swinton—"cistern," "reservoir," and "tank," all of which were applicable to the steel-like structure of the machines in the earlier stages of manufacture. Because it was less clumsy and monosyllabic the name "tank" was decided on.

******

(2) That this Committee shall be responsible for the supply of caterpillar machine-gun destroyers or land cruisers of the approved type; complete in every respect for action, including both primary and secondary armament. That it shall receive instructions as to supply and design direct from the General Staff, War Office, the necessary financial arrangements being made by the Accounting Officer, War Office.

(3) That, in order to enable the committee to carry out its work with the maximum of despatch and minimum of reference, it shall have full power to place orders, and to correspond direct with any Government department concerned. To be in a position to do this, it should have placed to its credit, as soon as its work commences, a sum equivalent to the estimated cost of fifty machines, which sum should be increased later if necessary by any further amount required to carry out the programme of construction approved by the General Staff. The committee should also be authorised to incur any necessary expenditure in connection with experimental work, engagement of staff, travelling and other incidental expenses during the progress of the work.

(4) That as the machines are turned out and equipped, they shall be handed over to the War Office for the purpose of training the personnel to man them.

(5) That the Committee be reconstituted with Lieutenant A. G. Stern as chairman.

(6) That since the officers of the R.N.A.S. will cease to belong to that service as soon as the 'Tank Supply Committee' is constituted, arrangements shall be made now for their payment from the same source that will bear the cost of constructing the

land cruisers and for their appointment as military officers with rank suitable to the importance of their duties.

The experimental machine was completed towards the end of 1915 and its preliminary trials gave most promising results.

On January 30, 1916, Mr. D'Eyncourt, as head of the "Admiralty Committee," entrusted with the design and manufacture of the trial machine, wrote to Lord Kitchener and informed him that the machine was ready for his inspection and that it fulfilled all the conditions laid down by the War Office, *viz.*—that it could carry guns, destroy machine-guns, break through wire entanglements, and cross the enemy's trenches, whilst giving protection to its own crew. Mr. D'Eyncourt also recommended that a number should be ordered immediately to this model, without serious alteration, and that whilst these were being manufactured the design of a more formidable machine could be developed.

On February 2 the first official trial of the new machine was held at Hatfield and was witnessed by the Lords Commissioners of the Admiralty, Mr. Lloyd George, Mr. McKenna, and various representatives of the War Office and Ministry of Munitions. Following this trial G.H.Q., France, on February 8 signified their approval of the machine and asked that the Army might be supplied with a certain number.

Arising out of the Hatfield trial it was decided to form a small unit of the Machine-Gun Corps, to be called the "Heavy Section," and Lieutenant-Colonel Swinton was appointed to command it with his Headquarters in London, a training camp being first opened at Bisley and later on moved to Elveden near Thetford.

As the "Admiralty Committee," with the Director of Naval Construction as chairman, had finished their work and produced an actual machine complete in all respects and fulfilling all requirements, it was then decided that the Ministry of Munitions should take over the production of the machines. On February 10 the Army Council consequently addressed a letter to the Lords Commissioners of the Admiralty requesting them to convey "the very warm thanks of the Army Council to Mr. E. H. T. D'Eyncourt, C.B., Director of Naval Construction, and his Committee, for their work in evolving a machine for the use of the Army, and to Mr. W. A. Tritton and Lieutenant W. G. Wilson, R.N.A.S., for their work in design and construction."

Two days later, on February 12, the "Joint Committee" was dissolved and a new committee, closely following the lines laid down at the Conference held in the offices of the "Committee of Imperial Defence," was formed under the Ministry of Munitions, and known as the "Tank Supply Committee."

Chairman.

Lieutenant A. G. Stern, R.N.A.S., Director of Naval Constructions Committee.

Members.

E. H. T. D'Eyncourt, Esq., C.B., Director of Naval Construction.

Lieutenant-Colonel E. D. Swinton, D.S.O., R.E., Assistant Secretary, Committee of Imperial Defence.

Major G. L. Wheeler, R.A., Director of Artillery's Branch, War Office.

Lieutenant W. G. Wilson, R.N.A.S., Director of Naval Constructions Committee.

Lieutenant K. P. Symes, R.N.A.S., Director of Naval Constructions Committee.

P. Dale-Bussell, Esq., Director of Naval Constructions Committee, Contract Department, Admiralty.

Consultant.

Captain T. G. Tulloch, Ministry of Munitions.

On February 14, 1916, Mr. D'Eyncourt wrote the following letter, which we quote in full, to Lieutenant-Colonel W. S. Churchill, commanding 6th Royal Scots Fusiliers, B.E.F., France, whose initiative and foresight were the true parents of the tank as a war machine:

Dear Colonel Churchill,
It is with great pleasure that I am now able to report to you that the War Office have at last ordered 100 landships to the pattern which underwent most successful trials recently. Sir D. Haig sent some of his staff from the front.
Lord Kitchener and Robertson also came, and members of the Admiralty Board. The machine was complete in almost every detail and fulfils all the requirements finally given me by the War Office. The official tests of trenches, etc., were nothing to it, and finally we showed them how it could cross a 9 ft. gap after climbing a 4 ft. 6 in. high perpendicular parapet. Wire entanglements it goes through like a rhinoceros through a field of corn. It carries two 6-pounder guns in sponsons (a *naval* touch), and about 300 rounds; also, smaller machine-guns, and

is proof against machine-gun fire. It can be conveyed by rail (the sponsons and guns take off, making it lighter) and be ready for action very quickly.

The king came (February 8, 1915), and saw it and was greatly struck by its performance, as was everyone else; in fact, they were all astonished. It is capable of great development, but to get a sufficient number in time, I strongly urge ordering immediately a good many to the pattern which we know all about. As you are aware, it has taken much time and trouble to get the thing perfect, and a practical machine simple to make; we tried various types and did much experimental work. I am sorry it has taken so long, but pioneer work always takes time and no avoidable delay has taken place, though I begged them to order ten for training purposes two months ago. I have also had some difficulty in steering the scheme past the rocks of opposition and the more insidious shoals of apathy which are frequented by red herrings, which cross the main line of progress at frequent intervals.

The great thing now is to keep the whole matter secret and produce the machines altogether as a complete surprise. I have already put the manufacture in hand, under the *ægis* of the Minister of Munitions, who is very keen; the Admiralty is also allowing me to continue to carry on with the same Committee, but Stern is now Chairman.

I enclose photo. In appearance, it looks rather like a great antediluvian monster, especially when it comes out of boggy ground, which it traverses easily. The wheels behind form a rudder for steering a curve, and also ease the shock over banks, etc., but are not absolutely necessary, as it can steer and turn in its own length with the independent tracks.

E. H. T. D'Eyncourt.

Between its institution in February and the following August the "Tank Supply Committee" underwent certain slight changes of organisation, the distribution of its duties rightly tending more and more towards centralisation. Shortly after its formation a "Tank Supply Department" was created in the Ministry of Munitions to work with the "Tank Supply Committee." This Supply Department was concerned with and was responsible for the initial output of the tanks which figured in the Battle of the Somme.

On August 1, 1916, the following resolutions were come to by the "Tank Supply Committee," and agreed to by the Minister of Munitions:

That the 'Tank Supply Committee' should in future be named the 'Advisory Committee of the Tank Supply Department.'
That a Sub-Committee consisting of Mr. D'Eyncourt, Mr. Bussell, and the Chairman, should be appointed to decide in questions of design and policy.

On August 22, the Committee was dissolved on the ground that the organisation for Tank Supply must be assimilated to that of the other Departments of the Ministry of Munitions, and the outcome of this was the formation of the "Mechanical Warfare Supply Department," with Lieutenant Stern as Chairman. This department continued in existence from now on until the end of the war. Its powers were wide, embracing production, design, inspection and the supply of tanks, and its energy was unlimited.

Whilst all these changes were in progress the tanks were being produced, and the personnel assembled and trained, and on August 13, 1916, the first detachment of thirteen tanks, being the right half of "C" Company, left Thetford for France, to be followed on August 22 by twelve tanks to complete the complement of "C" Company. On August 25 the right half of "D" Company entrained at Thetford for France, and on August 30 the remainder of the company followed. Tanks on arrival in France were transported to Yvrench, near Abbeville, where a training centre had been established under the command of Lieutenant-Colonel Brough, who had proceeded to France on August 3, to make the necessary arrangements. On September 4, Colonel Brough, having organised the training, returned to England, and Lieutenant-Colonel Bradley took over command of the Heavy Section.

It was now decided by G.H.Q. that tanks should take part in the next great attack in the Somme battle on September 15, so, on the 7th, 8th, and 9th of this month, "C" and "D" Companies moved to the forward area, and established their headquarters at the Loop, a railway centre not far from the village of Bray-sur-Somme.

Chapter 3
Mechanical Characteristics of Tanks

The following very brief account of the mechanical characteristics of tanks, it is hoped, will prove sufficiently simple and complete to give to the non-mechanical reader some idea of the tank as a machine.

The Mark I Tank (see frontis).

The first British tank made, and to be used, was the heavy machine, already described in the previous chapter, the Mark I tank, the general outline of which remained the standard design for the hulls of all British heavy machines up to the end of the war. As will be shown later, many mechanical improvements, making for higher efficiency and greater simplicity of control, were introduced from time to time, but the fact remains that the profile of the Mark V tank of 1918 was to all intents and purposes that of the Mark I of 1916, and surely this is a striking tribute to the genius of the designers who, without much previous data upon which to base their work, produced the parent weapon.

It is not proposed here to enter upon the general arrangement of the Mark I tank, but reference to two important points in design is of interest. The first is that this machine was fitted with a "tail," consisting of a pair of heavy large-diameter wheels, mounted at the rear of the machine upon a carriage, which was pivoted to the hull in such a manner that the wheels were free to follow the varying contours of the ground. A number of strong springs normally kept the wheels bearing heavily upon the ground, whilst a hydraulic ram, operated by an oil pump driven from the engine, was intended to enable the carriage to be rocked upon its axis, in order to raise the wheels well clear of the ground on occasions when it was necessary to "swing" the tank.

The object of this "tail" device was to provide means of steering the machine and, to this end, the driver was provided with a steering wheel which, operating a wire cable over a bobbin or drum, altered the path of the "tail" wheels, and allowed the tank to be steered, under favourable conditions, through a circle having a diameter of about sixty yards. The disadvantages of this fitting far outweighed any virtues it may have possessed. Countless troubles were experienced with the ram and its pump; the wire steering cable was constantly stretching or slipping through the bobbin, thus affecting the "lock" of the tail wheels; the driver was subjected to great physical strain in overcoming the tremendous resistance offered by the road wheels; the whole device was very prone to be damaged by shell-fire in action.

Against these indictments should be recorded the fact that the possession of a "tail" enabled the Mark I tank successfully to span and cross a wider trench than the later "tailless" machines of the same dimensions could negotiate, owing to the fact that as the wheels supported the rear of the tank over the point of balance, the risk of "tail dive" was considerably reduced. However, at the close of the operations of 1916, all tanks were shorn of their tails, and no subsequent models were fitted with them.

The second point of interest regarding this early machine relates to its sponsons. These, on both the male and female machines (armed with full length 6-pounder and Vickers machine-guns respectively) were planted upon and bolted to the walls of the hull and, for entraining purposes, these had to be removed and carried upon special trolleys which could be towed behind the tanks. It will readily be seen that this arrangement involved a considerable amount of labour, and rendered the process of entraining an extremely lengthy one; this led to an improved form of sponson being produced for the Mark IV machine.

The chief outstanding weaknesses of the Mark I machine, disclosed during the first appearance in action, were:

That the engine was provided with no silencer, consequently the noise, sparks, and even flames, which proceeded from the open exhaust pipes, passing through the roof of the tank, constituted a grave danger during the latter stages of an approach march. Many ingenious tank crews fitted to their machines crude types of silencers made out of oil drums, or adopted the plan of damping out the sparks by using wet sacks in relays, or covering the exhaust pipe with clay and mud.

That the observation from the inside of the machine was bad, and efficient fire control was, therefore, impossible.

That the means provided for entering and leaving the tank were unsatisfactory, and, in the case of the female machine, speedy evacuation in the event of fire was difficult.

That the whole of the petrol supply was carried inside the machine, and in a vulnerable position—a circumstance which added to the risk of fire in the event of a hit in the petrol tank by armour-piercing bullet or shell.

Furthermore, gravity was the only means for transferring petrol from the main petrol tanks in the front of the machine to the carburettor, and, therefore, it frequently happened that when a tank "ditched" nose downwards, the petrol supply was cut off, and consequently the dangerous practice of "hand-feeding" had to be resorted to.

The Mark II and Mark III Tanks

These machines were produced in small numbers, and their difference from Mark I lay in various minor improvements, none of a radical nature.

The Mark IV Tank (Plate 1)

In 1917 this tank became the standard fighting machine of the Tank Corps, and it was used in battle throughout this year and the following. As already stated, in outline it corresponded so closely with the Mark I machine that a study here of the main features of this tank will serve generally as an illustration of what had taken place in tank development up to this date.

The machine was 26 ft. 5 in. long over all, whilst the width of the female over its sponsons was 10 ft. 6 in., and of the male, 13 ft. 6 in. The height of the machine was 8 ft. 2 in., and its weight, equipped, was 28 tons. The armament consisted, in the case of the male, of two 6-pounder guns and four machine-guns, and in that of the female of six machine-guns; it was fitted with a 105 h.p. Daimler 6-cylinder sleeve-valve engine which, at a later date, was replaced in a limited number by one of 125 h.p. This increased power was obtained by the use of aluminium pistons, twin carburettors, and by speeding up the engine.

Generally speaking, these engines gave very little trouble, although somewhat under-powered for the work they had to perform. They were, it may be added, particularly suitable from the maintenance point of view, owing to their "fool-proof" nature, due chiefly to the absence of the usual poppet-valve gear, with its attendant risk of maladjustment.

Power was transmitted from the engine flywheel, through a cone-type clutch and a flexible coupling, to a two-speed and reverse gear-box, known as the primary gear, this being under the direct control of the driver, who could thus obtain first and second speeds, or reverse, without other assistance.

The tail-shaft from the gear-box carried a worm which drove the crown wheel of a large reduction gear, this gear also serving as a differential to enable the track driving wheels to rotate at different speeds, as when steering the tank on its track brakes. A device was provided, under the driver's control, for locking the differential when it was desired to steer a dead-ahead course, or when negotiating a trench or other obstacle. With the differential locked, the gear became, so to speak, "solid" and obviated the risk of one of the tracks slipping in bad ground, a condition very apt to cause a tank to slip sideways into a trench and become ditched.

Some trouble was caused through breakages of this locking muff in the earlier days, but latterly the arrangement was considerably improved and strengthened.

The gear-box tail-shaft terminated in a brake drum, the band of which was operated by means of a pedal at the driver's foot. It may be of interest to point out here that the whole of the items so far referred to, *i.e.* engine, gear-box, and differential, formed the standard power unit of the pre-war Foster-Daimler tractor, and thus provided a known quantity around which the rest of the detail was designed. This greatly facilitated production.

On either side of the differential case projected cross-shafts, the outer ends of which were supported in bearings mounted upon the outside wall of the tank, and, between the inner and outer walls of the hull, two sliding pinions were carried on a splined portion of the cross-shaft, one pair of pinions on each of the right and left hand sides of the tank.

In describing the remainder of the transmission, it will suffice to deal only with one side of the machine, the detail on either side being identical.

The sliding pinions, already alluded to, were operated by means of short levers by two gearsmen, whose sole duty it was to assist the driver, who signalled to them his requirements from his seat in the front of the tank, the two gearsmen being accommodated towards the rear of the machine on seats placed over the primary gear-box. The sliding pinions were of two sizes, known as the high-speed and low-speed pinions, and immediately in their rear was mounted another pinion assembly, also carrying two gear-wheels of different dimensions, with which the sliding wheels could be engaged at will—in other words, on each side of the tank there existed what were known as secondary gear-boxes, each offering a choice of two speeds.

Thus, it will be seen that the whole arrangement provided a range of four speeds. Assuming the secondary gears to be at "low," the driver had the option of using either first or second speed by manipulating the control to the primary gear-box, whilst in order to obtain third or fourth (top) speed it was necessary for him to signal the gearsmen to alter their gears to "high," and to assist them in the process by a great deal of intelligent clutch work. It need hardly be pointed out that this arrangement was exceedingly clumsy, and often involved much loss of time and temper. It might also be mentioned here that the reverse gear, already alluded to, was considerably higher than the lowest forward speed, so that there was little possibility of driving backwards, clear of any obstruction which might have ditched the tank.

Hand-operated brakes, under the control of the tank commander in the front of the tank, alongside the driver, were incorporated with the secondary gear-box. These brakes, by checking one or other track, enabled the tank to be steered in some measure with the differential unlocked, whilst, by locking the differential and placing, say, the right-hand secondary gear in "neutral," the machine could be swung to the right, practically upon its own axis, by applying the right-hand brake. To swing to the left, the right-hand secondary gear was engaged, the left-hand being placed in "neutral," the differential locked and the left-hand brake applied.

From the secondary gear-box a Coventry chain transmitted the power to an assembly, at the rear of the hull, which carried, on either side of the chain sprocket, two heavy pinion wheels, in constant mesh with the final sprocket wheels, which in turn, engaging with the links of the track plates, drove the hull along the track.

Each track was composed, normally, of ninety plates or road shoes, the separate plates being coupled together by means of links (two per plate) and link pins, the links themselves being recessed so as to engage with the driving wheels as shown above.

The weight of the machine was carried upon the track by means of rollers, whilst the track was supported on the top of the hull by skids or rails.

Adjustment of track was effected by the movement of an "idler" wheel, which guided the track over the nose of the hull.

Refinements to the transmission were introduced in the shape of guards to protect the driving chains from mud, and also means were provided to lubricate the secondary gear-wheels with oil. It is recalled that, prior to the introduction of the chain-guard, the inside deck of the tanks was often covered with a layer of liquid mud, several inches deep, carried in by the chains, and delivered through the secondary gears.

Petrol was supplied to the engine in the earlier days of the Mark IV machine by a pressure-fed system which gave a great deal of trouble, and, being also considered dangerous, was finally discarded in favour of the Autovac system, which sucked the fuel from the main supply in a tank outside of the machine and delivered it to the carburettor by gravity.

Cooling of the engine was primarily effected by a copper envelope radiator, which gave some trouble and was finally superseded by a tubular type.

An efficient silencer, with a long exhaust pipe carried right to the rear of the machine, considerably reduced engine noise and rendered the approach march a far less hazardous undertaking than was the case with the earlier models.

Sponsons were designed to collapse into the interior of the machine when necessary, and the cumbersome practice of detaching them from the hull came to an end. Short 6-pounder guns were introduced to render this change possible.

Detachable "spuds," to provide a grip for the tracks on difficult soil, were first introduced for this machine, as also was a highly efficient unditching gear. The latter consisted of a beam, rather longer than the overall width of the tank hull, which was fastened by clips and chains to each track, and, in passing round under the machine, actually took a purchase from the obstruction under the belly of the tank.

Detail improvements to give easier entrance and more rapid egress in case of emergency, as well as better and safer vision and fire control, were also introduced.

THE MARK V TANK (PLATE 5)

With the introduction of the Mark V tank, which represents the standard British heavy tank of today, great progress was made in all-round speed, ease of manoeuvre, radius of action, simplicity of control and feasibility of observation.

The dimensions and weight of this tank were approximately the same as those of the Mark IV, whilst the design of the hull still closely followed the lines of the original Mark I. Equipped with the 150 h.p. Ricardo 6-cylinder poppet-valved engine, specially designed for tank work, the advent of the Mark V machine called for the introduction of new courses of instruction for the personnel of the Corps, very few of either officers or other ranks having, at this time, any experience of the care and adjustment of the valve gear.

This Ricardo engine, of somewhat unorthodox design, was highly efficient and, with proper care and attention, gave very little trouble. From the engine, power was transmitted through a plate clutch in the flywheel to a four-speed gear-box, immediately in rear of which was the reverse gear, providing "reverse" *on all speeds*. The cross-shaft, incorporated with the reverse gear, carried at either end (in the same relative position as the secondary gears, explained in dealing with the Mark IV machine) an epicyclic gear. It is not within the scope of this chapter to describe this gear in detail, but it may be regarded as serving the double purpose of a reduction gear and clutch, combined in one unit.

From these epicyclic gears, the transmission of the drive through to the tracks followed the principle of the Mark IV machine, except that there was no second-line pinion assembly as in the secondary gear of the earlier tank, the Coventry chain on the Mark V passing direct from the single-unit epicyclic gear to the pinion assembly operating the track driving wheels.

All the items enumerated above were under the direct control of the driver, who was therefore enabled to perform, single-handed, all the operations which previously required the work of four men. Hand levers controlled the epicyclic gears, primary gear-box, and reverse gear, whilst the clutch and gear were foot-operated.

To steer the tank at any speed, the driver had merely to raise the epicyclic gear lever on the side on which he wished to turn. This had the effect of interrupting the drive to that track, so that, being driven by the remaining track, the machine would turn upon the "idle" side.

Where a sharp "swing" was necessary, application of the foot brake would automatically check the "idle" track, this being allowed for by means of a single compensating link motion with which the controls were interconnected.

The engine was petrol-fed by the Autovac system, as fitted to the later Mark IV machines. Cooling of the engine was effected by means of a tubular type radiator, the water therein being itself cooled by air drawn from outside the tank, through louvres in the left-hand wall of the hull, and finally expelled through similar louvres in the right-hand wall.

Further, the engine was completely enclosed in a sheet-iron casing, from which the hot foul air was exhausted through the roof of the tank by means of a Keith fan.

The Mark V armament corresponded with that of the Mark IV, whilst the sponsons were of similar design to those fitted to the latter type.

The absence of the large differential gear, as fitted on the earlier models, gave accommodation for a machine-gun in the rear wall of the tank, and also allowed for large entrance doors in the back portion of the roof.

A greatly improved type of rear cab was fitted, and thus provided excellent all-round vision, and also rendered possible the fitting of the unditching beam to the tracks from the inside of the machine. This was accomplished through the side flaps of the rear cab of the Mark V, whereas on previous models it had been necessary for members of the crew to expose themselves to hostile fire, in the event of the tank becoming ditched in action, as the beam could only be attached and detached from outside.

CHARACTERISTICS OF BRITISH TANKS USED DURING THE GREAT WAR

Characteristics.	Mark I. Male.	Mark I. Female.	Mark IV. Male.	Mark IV. Female.	Mark V. Male.	Mark V. Female.	Mark V star. Male.	Mark V star. Female.	Medium Mark A.	Gun-carrier.
Length with Tail	32′ 6″	32′ 6″	—	—	—	—	—	—	—	43′ 0″
Length without Tail	26′ 5″	26′ 5″	26′ 5″	26′ 5″	26′ 5″	26′ 5″	32′ 5″	32′ 5″	20′ 0″	30′ 0″
Width	13′ 9″	13′ 9″	13′ 6″	10′ 6″	13′ 6″	10′ 6″	13′ 6″	10′ 6″	8′ 7″	11′ 0″
Height	8′ 05″	8′ 05″	8′ 2″	8′ 2″	8′ 8″	8′ 8″	8′ 8″	8′ 8″	9′ 0″	9′ 4″
Weight, equipped	28 tons	27 tons	28 tons	27 tons	29 tons	28 tons	33 tons	32 tons	14 tons	34 tons
Crew	1 officer 7 O.R.	1 officer 7 O.R.	1 officer 7 O.R.	1 officer 7 O.R.	1 officer 7 O.R.	1 officer 7 O.R.	1 officer 7 O.R.	1 officer 7 O.R.	1 officer 2 O.R.	1 officer 3 O.R.
Armament	2 6-pdrs. and 4 Hotchkiss guns	5 M.G.s and 1 Hotchkiss gun	2 6-pdrs. and 4 Lewis guns	6 Lewis guns	2 6-pdrs. and 4 Hotchkiss guns	6 Hotchkiss guns	2 6-pdrs. and 4 Hotchkiss guns	6 Hotchkiss guns	4 Hotchkiss guns	1 Lewis gun
Engine	105 h.p. Daimler	105 h.p. Daimler	105 h.p. Daimler	105 h.p. Daimler	150 h.p. Ricardo	150 h.p. Ricardo	150 h.p. Ricardo	150 h.p. Ricardo	2 Tylor, 45 h.p. each	105 h.p. Daimler
Maximum Speed	3·7 m.p.h.	3·7 m.p.h.	3·7 m.p.h.	3·7 m.p.h.	4·6 m.p.h.	4·6 m.p.h.	4·0 m.p.h.	4·0 m.p.h.	8·3 m.p.h.	3·0 m.p.h.
Average Speed	2·0 m.p.h.	2·0 m.p.h.	2·0 m.p.h.	2·0 m.p.h.	3·0 m.p.h.	3·0 m.p.h.	2·5 m.p.h.	2·5 m.p.h.	5·0 m.p.h.	1·75 m.p.h.
Radius of Action	Hours 6·2, miles 12	Hours 6·2, miles 12	Hours 7·5, miles 15	Hours 7·5, miles 15	Hours 9·0, miles 25	Hours 9·0, miles 25	Hours 7·5, miles 18	Hours 7·5, miles 18	Hours 10·0 miles 40	Hours 7·5, miles 15
Spanning Power	11′ 6″	11′ 6″	10′ 0″	10′ 0″	10′ 0″	10′ 0″	14′ 0″	14′ 0″	7′ 0″	11′ 6″

NOTE.—(i) The Mark V star tank could carry twenty men in addition to its crew.
(ii) The Gun-carrier could carry 10 tons weight of stores.
(iii) Radii of Action are only approximate; they depend on the nature of the ground, efficiency of the crew, etc.

The Mark V One Star Tank (Plate 7).

The Mark V star machine was 6 ft. longer than the Mark V, and the weight of the male, equipped, was approximately 33 tons. There was no change in the nature of the armament, or in the number of the crew, which consisted of eight all told. In addition to the crew, the machine was capable of carrying twenty to twenty-five other troops and would cross a 14 ft. trench, as against 10 ft. for the Mark V.

The general mechanical arrangement of this tank corresponded with that of the Mark V, the same engine and transmission system being adopted, with the addition of a Cardan shaft between the flywheel and gear-box, which was rendered necessary by the additional length of the machine.

The Mark V star was relatively slow to manoeuvre, owing chiefly to the amount of track-bearing surface on the ground.

The Medium Mark A or "Whippet" Tank (Plate 3)

The Medium A tank, known also as the "Chaser" and "Whippet," was the British standard light-type machine, and it differed altogether from its heavier relatives. Its weight, equipped, was about 14 tons, whilst it was 20 ft. long, 9 ft. high, and 8 ft. 7 in. wide, carrying a crew of three. It could attain a maximum speed of about 8·3 m.p.h., and could span a trench approximately 7 ft. in width.

On this machine the tracks were not carried "overhead" as in the case of the heavy tanks, but the two trackways existed as such only, and formed the road members of what may be described as the chassis upon which the engine-room and fighting cab were mounted. There were no sponsons, and the tank was driven and fought from the cab at the rear of the machine, provision being allowed for an armament of three machine-guns.

Each machine was fitted with two 45 h.p. 4-cylinder Tyler engines with an Autovac petrol feed, and cooled by means of a tubular radiator provided with two fans driven by chains from each crankshaft.

The power of each engine was transmitted through separate cone clutches, leather flexible couplings, and four-speed and reverse gear-boxes, to a casing, at the rear of the machine, containing two worm gears.

The two worm-wheel shafts of these gears were in line, with their inner ends nearly touching, and each carrying the keyed-on half of a jawed coupling, one of which could be slid along at will, to engage with the other, thus locking the two shafts together.

One of the shafts carried a friction-clutch arrangement, designed to limit the power transmitted from one shaft to the other to about 12 h.p.

It will be seen, therefore, that either worm-wheel shaft could be driven independently by its own engine, or the two could be locked together so as to rotate at the same speed, driving the tank straight ahead, provided that there was not more than 12 h.p. difference between the developed powers of the engines. Extensions of each worm shaft carried a band brake, as well as a fan for forcing air into the cab of the machine.

Returning to the details of the transmission system, each cross-shaft from the worm case terminated in a "driving chain pinion shaft," the outer ends of which were supported by ball bearings mounted upon the sides of the track frames. The chain pinion carried by this shaft transmitted the drive, through a roller chain, to the final track-driving wheels, which, engaging with the slots in the track links, drove the tank along the track. Each track consisted normally of sixty-seven plates or shoes, and rollers served to support the weight of the tank upon the track, as well as to carry the track over the top of the trackway. Adjustment of the track was effected by movement of the front "idler" wheel as in the case of the heavier machines.

The Whippet tank called for particular skill in driving, and a great deal of practice was usually necessary to produce a really efficient driver. "Stalling" of one or both engines was a common occurrence during the earlier stages of training. Steering was effected by varying the speed of either engine, and the radius of movement was proportional to the difference in the speed of the two engines, this difference being controlled by means of a steering wheel connected to the two carburettor throttles, movement of the wheel producing acceleration of one engine and deceleration of the other simultaneously.

The Gun-carrying Tank (Plate 7)

Originally designed for carrying a 60-pounder gun or 6-in. howitzer and ammunition into action, these machines during 1918 were chiefly used for the transport of supplies across country. The engine, a 6-cylinder 105 h.p. Daimler, was placed right at the rear of the machine, and the general lay-out of the transmission corresponded with that of the Mark IV modified to suit the engine position, the primary and secondary gears, etc., being mounted forward of the engine in the case of this G.C. tank. The final drive to the track was at the rear, and exactly followed the Mark IV practice, whilst the track itself was carried on track frames, in this respect somewhat resembling the Medium A machine.

Four men were required to control the G.C. tank, the driver and brakesman being separately housed in two small independent cabs mounted one over each track towards the front of the machine, whilst the secondary gearsmen travelled in the body of the machine.

A system of signalling by signs from driver to other members of the crew was adopted.

Situated between the inner walls of the hull at the front of the tank was a "skid" or platform which could be drawn out, and its front lowered to the ground, forming an inclined runway up which the gun was hauled, by means of a winding gear operated from the engine, to its travelling position on the machine.

Drums for carrying ammunition for the guns were supported on platforms over the tracks immediately in rear of the two control cabs.

The first G.C. tanks were fitted with "tails," similar to those on the Mark I machines, but these were later on discarded.

The above includes the brief mechanical summary of the various types of British tanks used during the Great War, and though, undoubtedly, the future will bring with it many improvements and may radically change the whole form of the present-day tank, it is doubted if ever, in the whole history of mechanics—let alone warfare, a novel machine has been produced which has proved so efficient on first use and required in the long run of two years of war so few changes.

CHAPTER 4

The Mark I Tank and Its Tactics

The Mark I tank was the direct produce of the experimental machine which was officially tested on February 2, 1916. It may be defined as "a mechanically-propelled cross-country armoured battery," the maximum thickness of its armour being 12 mm. (The sponsons of the Mark I were only 10 mm. armour and not proof against A.P. bullets.)

The main tactical characteristics of all tanks may be placed under the headings of—mobility, security, and offensive power, and as regards the Mark I machine the following is a general description of these characteristics:

(1) *Mobility.*—The Mark I tank could move over flat ground at 100 to 120 yards a minute, over ground intersected by trenches at 30 to 40 yards a minute, and at night time at 15 yards a minute. It could cross

all forms of wire entanglements, crushing down two paths through them which were passable by two single files of infantry. It could span a trench 11 ft. 6 in. wide, surmount an obstacle 5 ft. high, and climb a slope of 1 over 2.

(2) *Security.*—The Mark I tank was proof against ordinary bullets, shrapnel, and most shell-splinters.

(3) *Offensive Power.*—Mark I tanks were divided into two categories: male and female. The former carried an armament of two 6-pounder guns and four Hotchkiss machine-guns, the latter of five Vickers and one Hotchkiss machine-guns. The normal amount of ammunition carried was for males 200 rounds and 10,000 rounds S.A.A., and for females 12,000 rounds S.A.A.

The chief limitations of the tank are connected with its mobility. For the Mark I type these limitations were as follows:

Its circuit in action was about 12 miles, and the fighting endurance of its crew 8 to 12 hours. It was not suited for traversing swamps, thick woods, streams with marshy banks, or deep sunken roads. It could be expected to cross shelled dry ground at a slow pace, but should this ground become sodden with rain it would find difficulty in doing so, and might frequently become ditched.

A tactical paper on the employment of this machine was put forward officially in February 1916 by Colonel Swinton, entitled "Notes on the Employment of 'Tanks.'" This document is of special interest as it is the first tactical note published on the use of tanks. The following are certain extracts taken from it:

> "The use made by the Germans of machine-guns and wire entanglements—a combination which has such power to check the advance of infantry—has in reply brought about the evolution of the 'caterpillar bullet-proof climbing motor, or tank,' a machine designed for the express purpose of assisting attacking infantry by crossing the defences, breaking through the obstacles, and of disposing of the machine-guns. It is primarily a machine-gun destroyer, which can be employed as an auxiliary to an infantry assault...."

> "Hostile machine-guns which it is impossible to crush (*i.e.* by running over them) will be attacked by gunfire. It is specially for the purpose of dealing with these weapons ensconced in houses, cellars, amongst ruins, in haystacks, or in other con-

cealed positions behind the enemy's front line, where they may not be knocked out by our artillery, and whence they can stop our infantry advances, that the tanks carry guns. Being covered with bullet-proof protection, and therefore to a great extent immune from the hostile machine-guns, they can approach sufficiently close to locate the latter, and pour in shell at point-blank range...."

"As ... it is proposed that the tanks should accompany the infantry," they should carry forward the following signalling apparatus, "small wireless sets ... an apparatus for laying a field telephone cable either on the surface of the ground or possibly buried 12 in. deep ..." also visual signalling apparatus and smoke rockets.

"The tanks will be destroyed by a direct hit of any type of howitzer shell. They will probably be put out of action by all except the most glancing hits of high-explosive shell fired by field-guns.... They may also be blown up by mines or land-mines...."

"Since the chance of success of an attack by tanks lies almost entirely in its novelty, and in the element of surprise, it is obvious that no repetition of it will have the same opportunity of succeeding as the first unexpected effort. It follows, therefore, that these machines *should not be used in driblets* (for instance, as they may be produced), but that the fact of their existence should be kept as secret as possible until the whole are ready to be launched, together with the infantry assault, in one great combined operation...."

"The sector of front where the machines can best operate should be carefully chosen to comply with their limitations, *i.e.* their inability to cross canals, rivers, deep railway cuttings with steep sides, woods and orchards...."

Tanks should remain at the position of assembly "sufficiently long for the crews to reconnoitre, ease and mark out the routes up to the points where they will actually cross the front defences, and to learn all that can be discovered of the German front-line trenches, and the defence zone behind it over which they have to advance...."

"The tanks, it is thought, should move forward together, say, by rocket signal, sweeping the enemy's first-line parapet with machine-gun fire; and after they have proceeded some three-quarters of the way across 'No Man's Land,' and have succeeded in attracting to themselves the fire of the German infantry and machine-guns in the front line, the assaulting infantry should charge forward so as to reach the German defences soon after the tanks have climbed the parapet and begun to enfilade the trenches...."

"... unless expectations are falsified, if the machines accompany the assaulting infantry, moving with it, or just ahead of it ... both will be across the enemy's front line and on their way to the second before the curtain of fire descends, and the latter will be behind them. It is hoped similarly that, owing to the prevention of the usual checks to the advance, which the action of the tanks will ensure, by the time the German gunners shorten the range in order to provide a second curtain in front of their second line, our assault will have already swept beyond the line.

"The above anticipations are admittedly sanguine; but if the tanks are employed and are successful, it is thought that they will enable the assault to maintain most of its starting momentum, and *break through the German position quickly,*" a condition which up to the present it has not been possible to attain, "even after immense sacrifice of life.".... "Not only, however, does it seem that the tanks will confer the power to force successive comparatively unbattered defensive lines, but ... the more speedy and uninterrupted their advance, the greater the chance of their surviving sufficiently long to do this. It is possible, therefore, that an effort to break right through the enemy's defensive zone in one day may now be contemplated as a feasible proposition.... This being the case, it appears that when the tanks are used the contingency of such an extended bound forward being made should be most carefully legislated for in the way of preparation to send forward reinforcements, guns, ammunition, and supplies...."

"The necessity for the co-ordination of all arms to work together in the offensive generally requires no remarks here, but the desirability of the specially careful consideration of the subject in the case of an operation by tanks, requires some emphasis, since the orchestration of the attack will be complicated by the introduction of a new instrument and one which somewhat alters the chain of interdependence of all. A recapitulation of this chain will make the matter clear. The tanks cannot win battles by themselves. They are purely auxiliary to the infantry, and are intended to sweep away the obstructions which have hitherto stopped the advance of our infantry beyond the German first line, and cannot with certainty be disposed of by shell fire. It follows, therefore, that the progress of the attack, which depends on the advance of the infantry, depends on the activity and preservation in action of the tanks.

"The weapons by which the tanks are most likely to be put out of action are the enemy guns. The only means by which we can at the earlier stages of an attack reduce the activity of the enemy's guns, are by our own artillery fire or by dropping bombs on them from the air.

"It follows, therefore, that in order to help our infantry in any operation in which tanks take part . . . the principal object of our guns should not be to endeavour to damage the German machine-guns, earthworks, and wire, behind the enemy's first line, a task they cannot with certainty carry out, and which the tanks are specially designed to perform. It should be to endeavour to help the infantry by helping the tanks, *i.e.* by concentrating as heavy a counter-fire as possible on the enemy's main artillery position, and on any field or other light guns whose situation behind the first line is known. . . ."

"In order to increase the confusion which, it is hoped will be caused amongst the enemy by an attack by tanks, and to assist in concealing the exact nature and the progress of these machines, it would be of advantage if their advance were heralded by clouds of smoke. . . ."

The above quotations need no comment, and if comment is to be sought for, the most suitable places to seek it are the battles in which tanks eventually took part, for in these, and the great number of lesser actions, some eighty-five in all, it will be found that not only were Colonel Swinton's speculations, made seven months before the first tank crossed "No Man's Land," not mere "flights of imagination," but "solid facts," the value of which these battles have proved again and again.

CHAPTER 5
The Battles of the Somme and Ancre

On July 1, 1916, the Battle of the Somme opened with a successful advance on the British right between Maricourt and Ovillers, and a check on the British left between Ovillers and Gommecourt. From that day on to the commencement of the Battle of the Ancre, in November, no further attempt was made to push forward the British left, all available troops being required to maintain the forward movement of the right flank.

The ground which separates the Rivers Somme and Ancre is split up into valleys by pronounced ridges, most of which form natural lines of defence for an enemy and could, in 1916, only be stormed after having been subjected to a heavy artillery bombardment. The ground had consequently become severely "crumped" in places; but as the weather, up to September 15, had been fine and dry, it offered no insuperable difficulty to the movement of the tanks, which were allotted to the Fourth and Reserve Armies as follows:

Fourth Army, XIVth Corps	"C" Company (less 1 Section)	17 tanks
,, ,, XVth ,,	"D" Company (less 1 Section)	17 ,,
,, ,, IIIrd ,,	1 Section "D" Company	8 ,,
Reserve Army	1 Section "C" Company	7 ,,
In G.H.Q. Reserve (all mechanically unfit)		10 ,,

On September 11 operation orders were received from the Fourth Army, and on the 13th a conference was held, at which Lieutenant-Colonel Bradley attended, to discuss the forthcoming attack. During the 14th "A" Company arrived at Yvrench, and at 4.30 p.m. on that day the headquarters of "C" Company moved to the Briquetterie near Trones Wood, and the headquarters of "D" Company to Green Dump.

The frontage of the Fourth Army attack extended between the Combles ravine and Martinpuich, the intention being to break through the enemy's defensive system and occupy Morval, Les Bœufs, Gueudecourt, and Flers. Simultaneously with this attack the Reserve Army was to attack on the left of the IIIrd Corps, and the French on the right of the XIVth Corps. The attack was to be pushed with the utmost vigour, and was to be followed by the advance of the Cavalry Corps, which was to seize the high ground about Rocquigny-Villers au Flos-Riencourt-lez-Bapaume.

The general idea governing tank movements, on this the first occasion of their use, was that they should be employed in sub-sections of two or three machines against "strong points." Considerable apprehension existed as to the likelihood on the one hand of tanks, if they started too soon, drawing prematurely the enemy's fire, and on the other of their reaching their objective too late to be an assistance to the infantry. It was finally decided that they should start in sufficient time to reach the first objective five minutes before the infantry got there, and thus risk drawing hostile fire. Our own artillery barrages, stationary and creeping, were to be brought down at zero, leaving lanes free from fire through which the tanks were to advance.

The tanks moved up from their positions of assembly to their starting-points during the night of September 14–15. Of the forty-nine machines allotted for the attack, thirty-two reached their starting-points in time for the battle, the remainder failing to arrive through becoming ditched on the way, or breaking down through mechanical trouble.

The tanks working with the Reserve Army and the IIIrd and XIVth Corps were not a great success; the operations of those with the XVth Corps were as follows:

The tanks allotted to this corps assembled on the night of September 13–14 at the Green Dump, where the machines were tuned up for battle, and where stores of petrol and oil had been collected. On the night of the 14–15th the tanks moved up to their starting-points round Delville Wood. Every tank was given the route it had to follow, and the time it was to leave the starting-point; this was in most cases about half an hour before zero (dawn), and was intended to be arranged so that the tanks should reach the German trenches a few minutes ahead of our own infantry. Briefly, the orders were for eight tanks to advance on the west of Flers, and six on the east of that village, their destination being Gueudecourt and the sunken road to the west of it. The tanks were to attack all strong points on their routes, and to assist the infantry whenever held up.

Of the seventeen tanks which moved off, twelve reached their starting-points; eleven of these crossed the German trenches and did useful work. One in particular gave great assistance where the attacking infantry were held up in front of the Flers line by wire and machine-gun fire; the tank commander placed his machine astride the trench and enfiladed it; the tank then travelled along behind the trench, and 300 Germans surrendered and were taken prisoners. Another tank entered Gueudecourt, attacked a German battery and destroyed one 77 mm. field-gun with its 6-pounders; the tank was then hit by a shell and caught fire; only two of its crew got back to our lines.

This attack on September 15, from the point of view of tank operations, was not a great success. Of the forty-nine tanks employed, only thirty-two reached their starting-points; nine pushed ahead of the infantry and caused considerable loss to the enemy, and nine others, though they never caught up with the infantry, did good work in clearing up points where the enemy was still holding out. Of the remaining fourteen, nine broke down from mechanical trouble, and five became ditched.

The casualties amongst the tank personnel were insignificant. Of the machines ten were hit in action and temporarily rendered useless, and seven were slightly damaged, but not sufficiently so as to prevent them returning in safety.

The next occasion upon which tanks were used was during the attacks of September 25 and 26, five being allotted to the Fourth Army, and eight to the Reserve Army. Of these thirteen tanks nine stuck in shell-holes, two worked their way into Thiepval, and after rendering assistance to the infantry met a similar fate, and one, working with the XVth Corps, carried out the first "star" turn in the history of tank tactics, which in the report of the XVth Corps is described as follows:

> On September 25, the 64th Brigade, 21st Division, attack on Gird trench was hung up and unable to make any progress. A footing had been obtained in Gird trench at N.32 d.9.1, (map reference), and our troops held the trench from N.26 c.4.5, northwards. Between these two points there remained approximately 1,500 yards of trench, very strongly held by Germans, well wired, the wire not having been cut. Arrangements were made for a tank (female) to move up from here for an attack next morning. The tank arrived at 6.30 a.m. followed by bombers. It started moving south-eastwards along the Gird trench, firing its machine-guns. As the trench gradually fell into our hands, strong points were made in it by two companies of infantry, which were following in the rear for that purpose.
>
> No difficulty was experienced. The enemy surrendered freely as the tank moved down the trench. They were unable to escape owing to our holding the trench at the southern end at N.32 d.9.1. By 8.30 a.m. the whole length of the trench had been cleared, and the 15th Durham Light Infantry moved over

the open and took over the captured trench. The infantry then advanced to their final objective, when the tank rendered very valuable assistance. The tank finally ran short of petrol south-east of Gueudecourt. In the capture of the Gird trench, eight officers and 362 other ranks were made prisoners, and a great many Germans were killed. Our casualties only amounted to five. Nearly 1,500 yards of trench were captured in less than an hour. What would have proved a very difficult operation, involving probably heavy losses, was taken with greatest ease entirely owing to the assistance rendered by the tank.

The last occasion upon which tanks were used during 1916 was on November 13 and 14, in the Battle of the Ancre, which completed the Somme operations for the year. Heavy rain had fallen, and the difficult ground along the River Ancre had been converted into a morass of mud. For this attack complete tank preparations were made, reconnaissances were carried out, and a tankodrome (Tank Park) was established at Acheux.

On account of the bad weather the original plan, namely, to use twenty tanks, was abandoned, and a much more modest scheme was evolved. Three tanks were to operate with the 39th Division opposite St. Pierre Divion. On November 13 these moved forward, and eventually all three stuck in the mud. North of the River Ancre two tanks were sent against Beaumont Hamel; these also became ditched. Next morning three more tanks were sent out to clear up a strong point just south of the last-named village. One of these was hit by a shell, and the remaining two, on reaching the German front line, became ditched. These two tanks were, however, able to bring their 6-pounders and machine-guns to bear on the strong point, and their fire proved so effective that after a short time the Germans holding it surrendered, and 400 prisoners were rounded up by the tank crews—2 officers and 14 other ranks.

From the point of view of the general observer it might be said that, except for one or two small and brilliant operations, the tank during the Battle of the Somme had not proved its value. The general observer, however, is seldom the best judge, and when the actual conditions under which tanks were used, during the autumn of 1916, are weighed and the lessons sorted, history's verdict, it is thought, will be, that they had so far proved their value that September 15, 1916, will in future be noted not so much for the successes gained on that day, but as the birthday of a new epoch in the history of war.

What were these lessons?

(1) That the machine in principle was absolutely sound, and that all it required were certain mechanical improvements.

(2) That it had not been given a fair trial. It had been constructed for good going and fine weather; it had been, unavoidably, used on pulverised soil, often converted by rain into a pudding of mud.

(3) That, on account of the secrecy it was necessary to maintain, commanders had little or no conception of the tactics to apply to its use.

(4) That sufficient time had not been obtainable wherein to give the crews a thorough and careful training.

(5) That tank operations require the most careful preparation and minute reconnaissances in order to render them successful.

(6) That tanks require leading and controlling in battle, and consequently that a complete system of communication is essential.

(7) That tanks, like every other arm, require a separate supply organisation to maintain them whilst fighting.

(8) That tanks draw away much fire from the infantry, and have as great an encouraging effect on our own troops as they have a demoralising one on the enemy's.

These are the main lessons which were learnt from the tank operations which took place during the battles of the Somme and the Ancre, and the mere fact of having learnt them justifies the employment of tanks during these operations. Further it must be remembered that, whatever tests are carried out under peace conditions, the only true test of efficiency is war, consequently the final test a machine or weapon should get is its first battle, and until this test has been undergone, no guarantee can be given of its real worth, and no certain deductions can be made as to its future improvement.

CHAPTER 6

The Growth of the Tank Corps Organisation

The word "Reorganisation" is a word which will never be forgotten by any member of the Tank Corps Headquarters Staff; it was their one persistent companion for over two years. It dogged their steps through all seasons, over training areas and battlefields in sleuthhound fashion from the earliest days; and its pace was never stronger or its tongue more noisy than when, on November 11, 1918, it was temporarily shaken off with the armistice. Depressing as this perpetual change often was, reorganisation is, nevertheless, an extremely healthy sign, for it shows that the Tank Corps, a young formation, was not afraid to grow, and that it refused to stand still; and, when all is said and done, should not every organisation be dynamic, should not it move with the times, expand, grow, and absorb difficulties rather than push them aside or ignore them? Whatever, in the eyes of others, the Tank Corps may have been, throughout the Great War it was an intensely virile formation.

In this chapter the organisation and reorganisation of the Tank Corps, first known as the "Heavy Section," and later as the "Heavy Branch" of the Machine Gun Corps, will be dealt with in its entirety; for unless we lay this spectre in a chapter of its own it will never leave us in peace, but will haunt our steps right through this brief history, as was its wont when the incidents now related were taking form in France and England.

In June 1916 the Heavy Section Machine Gun Corps was organised in six companies—A, B, C, D, E, and F. Each company consisted of four sections, each of six tanks with one spare tank per company—in all twenty-five machines, thus absorbing the 150 machines ordered. (The original order was for 100, this was later on increased to 150). Each section consisted of 3 male and 3 female tanks, subdivided into three sub-sections of 1 male and 1 female each.

The crew of a tank was 1 officer and 7 other ranks, the total personnel of a section being 6 officers and 43 other ranks. For every two companies was provided a Quartermaster's establishment of 1 officer and 4 other ranks, and a workshop of 3 officers and 50 other ranks.

A few days after the Heavy Section had made its debut on the battlefield of the Somme, a suggestion was put forward to organise it on the lines of the Royal Flying Corps, which, eventually, in the main was adopted. This was undoubtedly a sound suggestion, as every new weapon requires an organisation of its own to nurse it through its infancy.

On September 29, Lieutenant-Colonel H. J. Elles, D.S.O., who, as we have seen, first came into contact with tanks in January 1916, was appointed Colonel Commanding the Heavy Section in France, and on the same day that his appointment was sanctioned it was decided that 1,000 tanks should be built, and that certain improvements in the existing design of machine should be introduced. At this time the Headquarters of the Heavy Section were located in one small hut in the centre of the square of the village of Beauquesne, and as this village was not considered suitable for a permanent Headquarters, Bermicourt was selected instead—a small village just north of the Hesdin-St.-Pol road. At this village the Headquarters remained until the end of the war, expanding from three Nissen huts to many acres of buildings.

On October 8 a provisional establishment for the Headquarters was approved. It consisted of—a commander (colonel), one brigade major, one D.A.A. and Q.M.G., one staff captain, and one intelligence officer. These appointments were filled by the following officers: Colonel H. J. Elles, Captain G. le Q. Martel, Captain T. J. Uzielli, Captain H. J. Tapper, and Captain F. E. Hotblack.

At about this time it was proposed to form the Heavy Section into a Corps, giving it an Administrative Headquarters in England and a Fighting Headquarters in France, and of converting the four companies in France into four battalions, and raising five new battalions in England on the nuclei of the two remaining companies. Though the formation of the tank units into a Corps was not sanctioned at the time the other proposals came into force on October 20, Brigadier-General F. Gore Anley, D.S.O., being appointed Administrative Commander of the Tank Training Centre, Bovington Camp, Wool, in the place of Colonel Swinton, with Lieutenant-Colonel E. B. Mathew-Lannowe as his G.S.O.1. Under this organisation the 9 battalions were eventually to be formed into 3 brigades each of 3 battalions, a battalion consisting of 3 companies, each company of 4 fighting sections and a headquarters section. A fighting section consisted of 5 tanks and the headquarters section of 8. In all the battalion was, therefore, equipped with 72 machines.

On November 18, the day on which the approved establishments were issued, the companies, which had continued in the area of operations, were moved to the area round Bermicourt and, ceasing to exist as companies, became A, B, C, and D Battalions Heavy Branch Machine Gun Corps. They were located at the following villages:

A Battalion	Humières, Eclimeux, Bermicourt
B "	Sautrecourt, Pierremont, St. Martin-Eglise
C "	Erin, Tilly-Capelle
D "	Blangy

These battalions were eventually formed into the 1st and 2nd Tank

Brigades: the 1st Brigade, consisting of C and D Battalions, on January 30, 1917, under the command of Colonel C. D'A. B. S. Baker Carr, D.S.O.; and the 2nd Brigade, of A and B Battalions, on February 15, under that of Colonel A. Courage, M.C. Later, on April 27, in view of the expected arrival of two battalions from Wool, approval was given to the formation of the 3rd Brigade Headquarters under the command of Colonel J. Hardress Lloyd, D.S.O.

Meanwhile, in England, the whole question of future production was being strenuously dealt with by Lieutenant-Colonel Stern, who, on November 23, assembled a conference in London at which the future production of tanks was explained as follows:

That at the time of the conference there were 70 Mark I machines in France, and it was hoped to deliver improved types of this tank as follows: 50 Mark II tanks by January; 50 Mark III tanks by February 7; Mark IV tanks at the rate of 20 per week from February 7 to May 31. Further, that Mark V tanks would be available in August and September 1917, and that a new light tank, called Mark VI, would be ready for trial by Christmas 1917.

Unfortunately, on account of the difficulty of production and the constant changes demanded in design, the above programme never materialised, and though Mark II tanks were sent out to France, no Mark IV machines arrived there until after the Battle of Arras had been fought and won.

Early in the new year the battalions of the Heavy Branch underwent a further reorganisation: they were slightly reduced in size and the number of their machines was cut down from 72 to 60; each company, theoretically consisting of 20 tanks, was divided into 4 sections of 5 tanks each; for practical purposes, however, it was found that a section could not deal with more than 4 tanks, so the number of tanks was reduced to 48, of which 36 were earmarked as fighting and 12 as training machines.

In March 1917 General Anley was appointed Administrative Commander Heavy Branch Machine-Gun Corps with his headquarters in London, Brigadier-General W. Glasgow taking over the command of the Training Centre at Wool. In May he was succeeded by Major-General Sir John Capper, K.C.B., and the Tank Committee under his chairmanship was formed to systematise and strengthen co-operation between the Army and the Ministry of Munitions. On the 1st of this month, Colonel Elles was gazetted Brigadier-General Commanding the Heavy Branch in France.

The experiences gained during the Battle of Arras, in April 1917, resulted in proposals being put forward for the expansion of the Heavy Branch from nine to eighteen battalions, nine to be equipped with heavy, and a similar number with medium machines. (The lighter form of tank was called "medium" because the French, by now, had produced the light Renault tank, see Plate 3).

These proposals mark an important stage in the development of the Heavy Branch and they were destined to be the subject of many discussions.

On June 28, the above expansion was authorised, and the personnel for new units was assembled at the Training Centre at Wool. A month later, however, the call for manpower became so urgent that the expansion of the Heavy Branch had to be suspended. It was on the 28th of this month that the Heavy Branch became known as the Tank Corps.

During the following months, August and September, the question of the Tank Corps expansion was held in abeyance. On October 6 it was once again revived, and a revised establishment for the contemplated expansion to eighteen battalions was submitted. The outstanding feature of these establishments was the abolition of Battalion Workshops and the substitution of Brigade Workshops in their place. This resulted in a considerable economy of man-power, and was rendered possible by the higher training of the tank crews; each tank with its crew thus tended to become a self-contained unit.

On November 27 these establishments received official approval, and exactly one week later, on December 4, arising out of the overwhelming success gained by tanks at the Battle of Cambrai (November 20), two new organisations were put forward, the first known as the Lower, and the second as the Higher Establishments. The Lower Establishments were eventually decided upon, and they consisted in a revised edition of the former establishments with various additions, which the experiences gained at the Battle of Cambrai had shown to be necessary. These establishments, though made out, were never approved, and the German offensive in March 1918 found the Tank Corps still organised on the lines agreed upon in October.

In April, on account of the pressing needs for infantry reinforcements, the Tank Corps expansion was temporarily suspended, two of the three remaining battalions in England being reduced to cadre units, and the third converted into an Armoured Car Battalion. In July and August, the astonishing successes gained by tanks on various sectors of the Western Front once again brought forward the need of increasing the British tank battalions, and the suspension was removed, the two remaining battalions of the expansion of October 1917 proceeding to France in September 1918.

In January 1918, from the experience gained by now in the time necessary to carry through a reorganisation, proposals were put forward for 1919. These were eventually discussed at the Inter-Allied Tank Committee, an assembly of representatives of the various allied Tank Corps, which first met at Versailles in April. The German spring offensive, however, absorbed so much attention that it was not possible at the time to do more than work out, as a basis, the number of tanks required for a decisive tank attack the following year. As the position of the Allies in France stabilised the question first discussed at Versailles was in July retaken up, with the result that an expansion to thirty-four battalions was decided on and completely new establishments called for. In order to bring this work more closely under the War Office it was also decided, at about this time, to dissolve the Tank Directorate, first created in May 1917, and to replace it by a new sub-branch of the Directorate of Staff Duties. This change took place on August 1, when a new branch known as S.D.7 was added to the Directorate of Staff Duties at the War Office to deal with the administration of tanks generally, and the 1919 tank programme in particular.

At the same time the Tank Committee was abolished, its place being taken by the Tank Board, which was constituted as follows:

Major-General the Right Honourable J. E. B. Seely, C.B., C.M.G., D.S.O., M.P., President (Deputy Minister of Munitions).

Sir Eustace Tennyson D'Eyncourt, K.C.B., Vice-President (Director of Naval Construction).

Admiral Sir Reginald Bacon, K.C.B., K.C.V.O., D.S.O. (Controller Munitions Inventions).

Major-General Sir William Furse, K.C.B., D.S.O. (Master General of Ordnance, representing the Army Council).

Major-General E. D. Swinton, C.B., D.S.O.

Major-General H. J. Elles, C.B., D.S.O. (Commanding Tank Corps, France).

Lieutenant-Colonel Sir Albert Stern, K.B.E., C.M.G. (Commissioner Mechanical Warfare, Overseas and Allies Department).

Colonel J. F. C. Fuller, D.S.O. (D.D.S.D. Tanks: representing General

Staff, War Office).

 Mr. J. B. Maclean (Controller of Mechanical Warfare).

 Sir Percival Perry (Inspector of Mechanical Traction).

 Captain A. Earle, Secretary.

The constitution of the Board is interesting as it enabled expert naval, military, and industrial knowledge to be concentrated on the one subject—the application of naval tactics to land warfare. The work accomplished by this Board was considerable, it was carried out in a high co-operative spirit and with great good-fellowship, and it would, undoubtedly, have proved a factor of no small importance in the complete destruction of the German armies in 1919, which was practically fore-ordained by a tank programme of some 6,000 machines, had the war continued.

September was a month of great activity at the Training Centre at Wool, and an extensive building programme was commenced under the direction of Brigadier-General E. B. Mathew-Lannowe, D.S.O., who had taken over the command of the Training Centre on August 1 from Brigadier-General W. Glasgow, C.M.G.

On October 22 the new establishments were received at the War Office, and were approved of and returned to G.H.Q. four days later. Considering that these establishments covered ninety-six pages of typed foolscap it may be claimed that the last reorganisation the Tank Corps experienced during the Great War was carried through in record time.

CHAPTER 7
Tank "Esprit De Corps"

The first "Instructions on Training" were issued to battalions of the Heavy Branch towards the end of December 1916. They are of some interest, as the *esprit de corps* and the efficiency of the entire formation was by degrees moulded on them.

> The object of all training is to create a *'corps d'élite,'* that is a body of men who are not only capable of helping to win this war, but are determined to do so. It cannot be emphasised too often that all training, at all times and in all places, must aim at the cultivation of the offensive spirit in all ranks. The requirements, therefore, are a high efficiency and a high moral.
>
> Efficiency depends on mental alertness and bodily fitness; the

first is produced by extensive knowledge and rapidity of thinking logically, the second by physical training, games, and the maintenance of health.

Moral depends on *esprit de corps* and *esprit de cocarde*; the first is produced by discipline, organisation and skill, the second by pride, smartness and prestige.

Efficient instructors and leaders are essential; indifferent ones must be ruthlessly weeded out. Officers must not content themselves with the teaching and knowledge they gain, but must supplement these by personal study and effort. Further, they must exercise their ingenuity in adapting the knowledge they have gained so that it may interest and expand the ideas of those they teach. In mental superiority and bodily vigour, they must be examples to their men.

As a general principle, officers and N.C.O.s, charged with the duty of instruction of troops, should adopt the following method: First the lesson is to be explained, secondly demonstrated, and finally carried out as an exercise.

Instruction *must* be interesting. As interest soon flags, subjects will be changed at short intervals, though the same movements must be frequently practised on different occasions.

Changes should be based on a system; thus, work which has required brain power should be followed by work entailing physical exertion, and *vice versa*. As physical training develops muscle on a definite system, so should mental training develop mind. It will not be easy to accomplish this unless schemes are carefully organised and thought out, and training is carried out according to a progressive programme.

Much time is often wasted by attempting long unrealistic movements and by prolonged drill. Three to four hours a day, divided into hourly or half-hourly periods, should be sufficient. Ten minutes' rest intervals should succeed each hour's work.

All work must be carried out at high pressure. Every exercise and movement should, if possible, be reduced to a precise drill. Games will be organised as a definite part of training.

Order is best cultivated by carrying out all work on a fixed plan. Order is the foundation of discipline. Small things like marching men always at attention to and from work, making them stand to attention before dismissing them, assist in cultivating steadiness and discipline. Each day should commence with a careful inspection of the billets and the men, or some similar formal parade. Strict march discipline to and from the training grounds must be insisted upon.

It is an essential part of training for war that the men are taught to care for themselves, so as to maintain their physical fitness. To this end the necessity for taking the most scrupulous care of their clothing, equipment and accoutrements will be explained to them.

The importance of obedience to orders will be impressed on all ranks and prevention of waste rigorously enforced.

Both in the case of officers and N.C.O.s special attention should be paid to the training of understudies for all positions and appointments.

The men must be brought to understand that on the skill they gain during training will depend their lives as well as the result of the battle. Instruction is not a matter of getting through a definite time, but of employing that time to the fullest advantage.

The training of the Heavy Branch was divided into the following categories: Brigade Training, Battalion Training, Schools, Courses of Instruction, Camps of Instruction, Lectures and Depot Training.

Brigade and battalion training were divided into two periods—individual training and collective training. As time was very limited, all individual training had to be completed by February 15, 1917.

> The object of individual training (to quote the *Instructions*) is twofold: first, to impart technical knowledge and skill; secondly, to cultivate general knowledge so as to enable all ranks to obtain the highest benefit from the schemes set in collective training. These latter in their turn are for the purpose of training units for battle. Individual training is the keynote of efficiency. On the thoroughness with which it is carried out rests the efficiency of the whole training.

The object of the collective training was:

> To apply, in conditions as near as possible to those which will be met with in battle, the detail learnt during individual training. This comprises:
>
> (1) Close co-operation with the other arms.
>
> (2) Rapidity of movement across ground in fighting formations.

(3) Selection of objectives with reference to the plan of operations.

During January and February all officers took part in a long indoor scheme which when completed formed a tactical and administrative basis for future operations, and all ranks were lectured to on discipline, *esprit de corps*, moral, and leadership.

Whilst the above work was in progress a Reinforcement Depot was formed, first at Humerœuil, later on it was moved to Erin, and eventually to Mers, near Le Treport. The Depot was the receiving station of all drafts arriving for the Tank Corps, whether from the Training Centre in England, or from units or hospitals in France. The duty of the Depot was to hold on its strength all reinforcements until fully trained, and when fully trained to continue refresher training until they were required to fill vacancies in the battalions.

Besides the Depot and the schools attached to it, two main schools—Gunnery and Tank Driving—were instituted in the Bermicourt area. In the early summer of 1917, the first was moved to the sea coast at Merlimont, and the second to Wailly, a village close to the zone devastated by the Germans during their retreat in the preceding February and March, which permitted of driving being carried out without damage to crops. This school remained at Wailly until January 1918, when, on account of the threatening German attack, it was moved to Aveluy near Albert. As it happened, Aveluy fell into the German hands towards the end of March 1918, whilst Wailly remained in ours until the end of the war.

Closely connected with the training of the men was the general administration of the Heavy Branch. It was fully recognised that the efficiency of all ranks depended to a great extent on the cheerfulness and comfort of their surroundings, and nothing was left undone, or at least unattempted, which could increase the men's happiness and health.

On January 1, 1917, baths and laundries were opened at Blangy. The arrangements first made enabled 450 men to bathe each day; this permitted of every man getting a bath once a week. Cinema theatres were also established at the Depot, and later on at Merlimont and elsewhere, being bought out of funds provided by the canteens' and supper bars. While at Erin a Rest Camp was formed to which those men who were temporarily incapacitated for work were sent to recuperate. This later institution was found so useful that in the summer of 1917 a seaside Rest Camp was established at Merlimont, the object of which was to provide rest and change of surroundings to men who had been in action, or whose health was impaired. This camp could accommodate 100 officers and 900 other ranks, and the period of rest there was usually limited to fourteen days.

An even more popular institution than the Merlimont Rest Camp was that of the Mobile Canteens: these consisted in lorries fitted to carry canteen stores; they formed the mechanical *vivandières* of the Tank Corps, following up units to within a mile or two of our front lines or pushing forward across the battlefield when a success had been gained. During the dark days of March and April 1918, they played a notable part in maintaining the *esprit de corps* of the battalions by providing comforts which would otherwise have been unobtainable. They also formed cheerful rallying-points where men could meet, eat, and chat, and then return to battle refreshed and still more determined to see it through for the honour of the corps to which they belonged and which, it may without boasting be said, always thought of their needs first and generally supplied them.

Chapter 8
Tank Tactics

The training of the Heavy Branch having been laid down, it was next necessary to discover and decide upon a common method of tactics, (at this time, January 1917, General Swinton's notes given in Chapter 4 were not known of at the Heavy Branch Headquarters), so that directly individual instruction had been completed collective training might be based on it; further, rumours were already afloat that the Heavy Branch might be called upon to take part in the spring offensive, so there was no time to be lost in deciding upon suitable methods and formations of attack. This was done early in February, when "Training Note No. 16," which will long be remembered by many in the Tank Corps, was issued.

Though experience is the only true test of a system of tactics, the foundations of the tactics suitable to any particular weapon are not based on experiences, but on the limitations of the weapon, that is on its powers, and on the fundamental principles of war. Further than this, if the weapon concerned is to be employed in co-operation with other weapons, the powers of these other weapons must also be considered, so that all the weapons to be employed may, so to speak, like a puzzle, be fitted together during battle to form one united picture.

In thinking out a tactics for tanks, the first factors to bear in mind are the powers of the machine, which may be summarise in three words: "penetration with security." Heretofore fronts had remained to all intents and purposes inviolable to direct infantry attacks; the tank was now going to break down this deadlock through its ability to cross wire and trenches under fire with far less risk than infantry could ever hope for. Mechanically, the machine was far from perfect, consequently, it was laid down, as a general rule, that never fewer than two tanks should operate together, and when possible not fewer than four.

From a military point of view the penetration of a line of defence does not simply mean passing straight through it, but cutting it in half, and then by moving outwards as well as forwards to push back and envelop the flanks thus created and so widen the base of operation to admit the movement forwards of reserves and supplies, and the movement backwards of casualties and tired troops. A man getting through a hedge first selects a weak spot (point of attack), he then forces his arms through the branches (penetration), and pushing them outwards (envelopment), forms a sufficiently large gap (base of operations) to permit of his body (army) passing easily through the hedge (enemy's defences).

The operation of penetration with tanks is just the same. Take a half section, two machines; this half section first penetrates the enemy's defences by crossing them (see diagram 7), then by moving outwards, say to the left, starts enlarging the base by driving the enemy towards A, and so makes a gap between the point of penetration and A, for the infantry to move through. As the enemy may, whilst the tanks are working towards A, seek refuge in his dugouts and "come to life" again after the tanks have passed by, it is necessary that the tanks should be followed by an infantry "mopping up" party which will bomb the dugouts and so render "coming to life" less frequent. As the bombing party has to work up the trench with the tank, it cannot hold the trench once it is cleared, consequently another party of infantry should follow the bombers, whose duty it is to garrison the trench on it being captured. We therefore find that even in the smallest tank attack two parties of infantry are required: in trench warfare these are known as "moppers up" and "support," and in field warfare as "firing line" and "supports." Frequently it is as well to add another party, a "reserve," so that some definite force of men may be held in hand to meet any unexpected event.

If instead of two tanks we use four, a much more effective opera-

Diagram 7

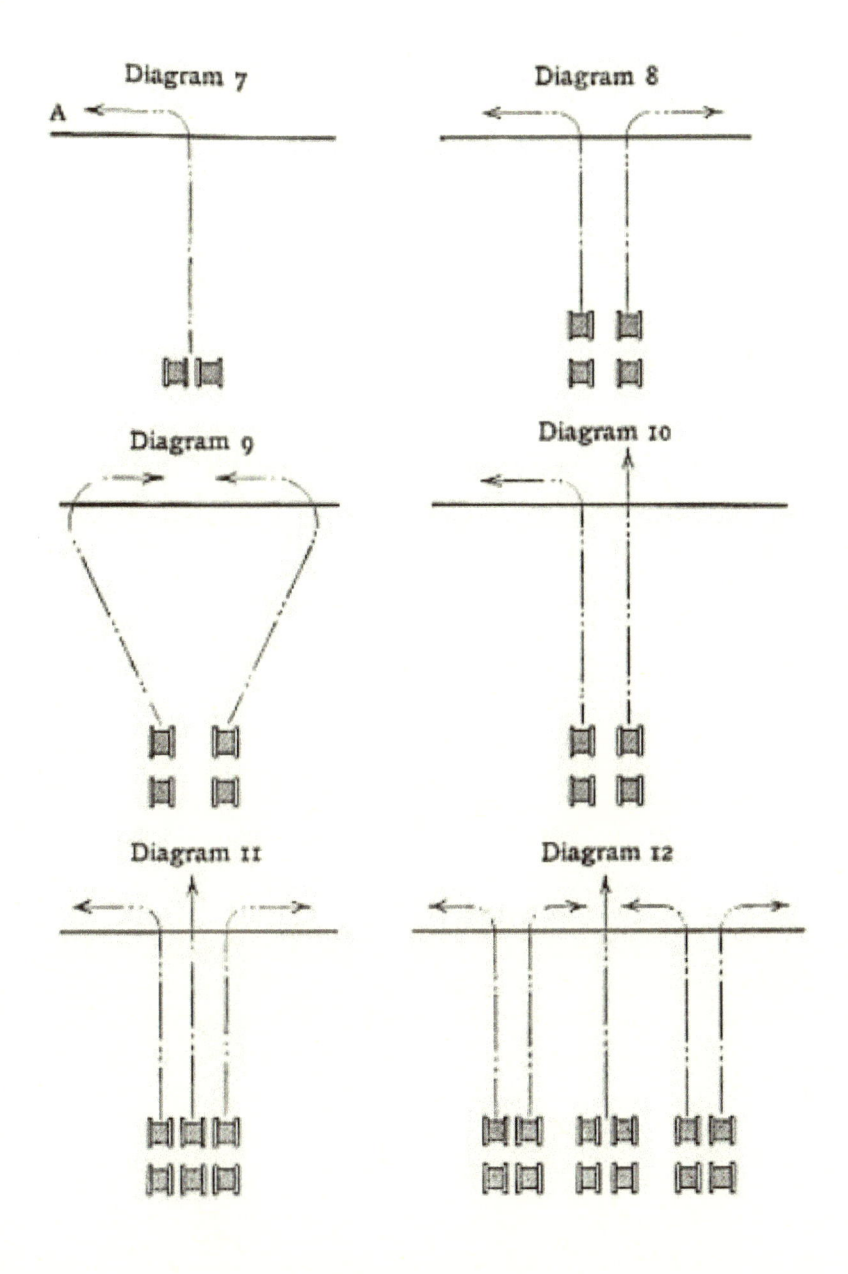

Diagram 8

Diagram 9

Diagram 10

Diagram 11

Diagram 12

tion may be carried out. The tanks can either penetrate at one place, and wheel outwards by pairs (see diagram 8), or by pairs penetrate at two separate points and wheel inwards, pinching on the centre (see diagram 9), or two can wheel to a flank and two proceed straight ahead (see diagram 10) and threaten the enemy's line of retreat. When this latter operation is contemplated it is as well to make use of at least six machines, better twelve, *i.e.* a complete company of tanks. If six machines are used, they normally should strike the enemy's line at approximately the same place; from there one half section should go straight forward and one to each flank, forming what has been called the "Trident formation" (see diagram 11). If twelve machines are employed, then each section of four tanks strikes the trench at a separate point, the centre section forging straight ahead and the flanking sections moving inwards and outwards as depicted in diagram 12.

Particular attention should be paid to the outward movements of the flanks, for, as the flanks of our own penetrating or attacking force are generally the most vulnerable points, if we can push forward offensive wings on these flanks we shall not only be protecting our own flanks from attack, by giving the enemy no time to attack in, but we shall be protecting our central line of advance as well. The force operating along this central line not only depends for its movement forward on the security of its flanks, but also on the size of the base of operations; the broader this base the more secure will it be, for the one thing an attacking army wishes to avoid is getting into a pocket on the interior of which all hostile fire is concentrated.

From the above elementary movements can be worked out a whole series of battle formations according to the various arms which are to be employed. The following three were those generally used by tanks from the Battle of Cambrai onwards:

(1) *An attack against trenches with an artillery barrage* (see diagram 13).—Three tanks in line at 100 to 200 yards' interval, followed by infantry in sections, each section forming an independent fighting unit advancing in single file and attacking in line, the whole forming one firing line. Behind this firing line should advance one tank and a certain number of infantry sections as a support. Reserves can be added as necessary.

(2) *An attack against trenches without an artillery barrage* (see diagram 14).—One tank in advance, followed at a distance of 100 to 150 yards by two others at 200 to 300 yards' interval, and one tank in support.

Diagram 13

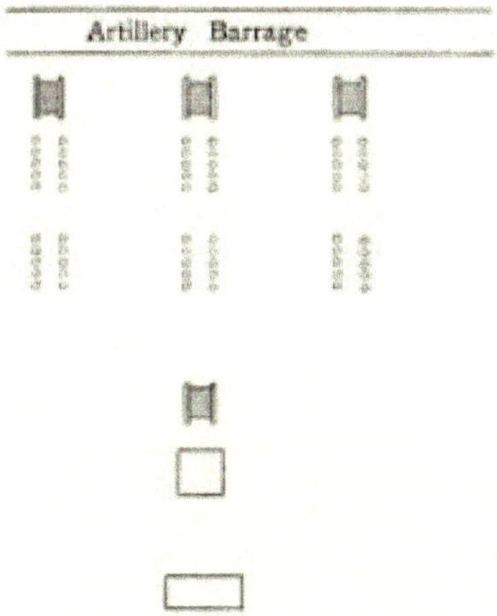

Artillery Barrage

Diagram 14

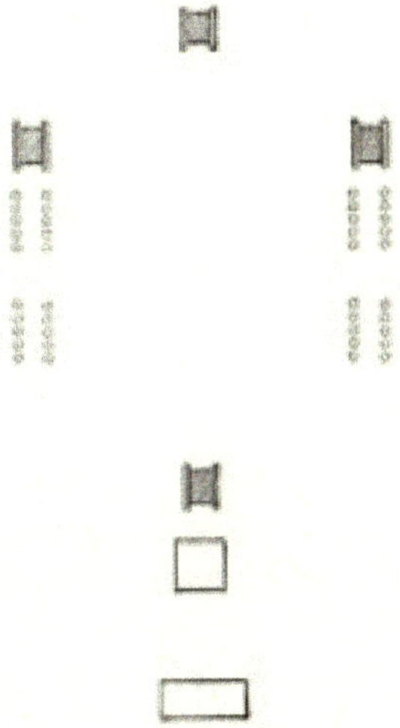

Diagram 15

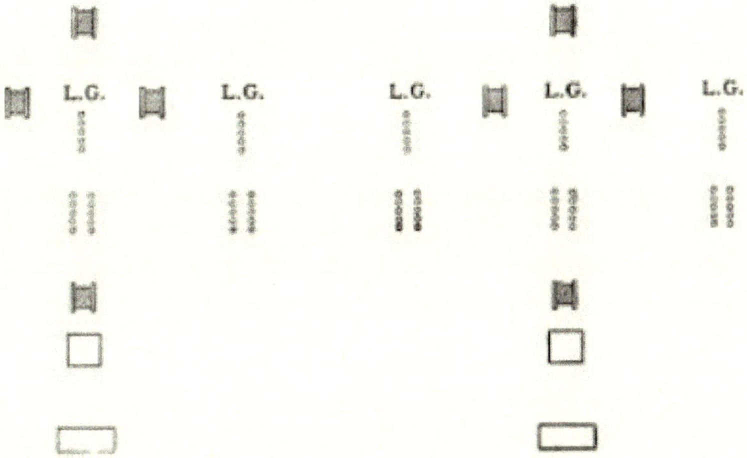

L.G. L.G. L.G. L.G. L.G.

The infantry should be disposed of as before. The advanced tank to a certain extent replaces the artillery barrage and acts as a scout to the two behind, which form part of the infantry firing line.

(3) *The field warfare attack* (see diagram 15).—In the field attack the action of the tanks must be adapted to circumstances. This action falls under three headings:

(*a*) Moving in front of the infantry firing line.

(*b*) Moving with the infantry firing line.

(*c*) Moving behind the infantry firing line.

When moving with the infantry firing line, which will generally be the most suitable formation to adopt, tank sections should form mobile strong points or bastions, which will not only reduce the number of infantry required for the firing line, but which will be able to bring oblique and cross fire to bear in front of the advancing infantry. In order to reduce the human target as much as possible without reducing fire effect, Lewis-gun sections should freely be used to cover by fire the intervals between tank sections. These Lewis-gun sections should be followed by rifle sections which, directly opposition is broken down by the tanks and the Lewis gunners, should move rapidly forward several hundred yards in front of the tank and infantry firing line, forming to it a protective screen of sharpshooters. This formation should then be maintained until the rifle sections get hung up, when the tank and Lewis-gun firing line should pass through them to renew the attack, the rifle sections forming up in support in rear. Curiously enough this formation resembles very closely that generally adopted by the Roman *Velites* and *Hastati* (riflemen), *Principes* (Lewis gunners), *Triarii* (tanks), and Napoleon's Light Infantry (riflemen), Infantry of the Line (Lewis gunners), Old Guard and Heavy Cavalry (tanks).

As an infantry attack depends on the following principles—the objective, the offensive, security, mass, economy of force, surprise, movement, and co-operation—so does a tank attack. The tank must know what it is after, it must act vigorously, it must be protected by artillery just like infantry, it must attack in mass, that is in strength and numbers, but not, necessarily, all in one place; it must surprise the enemy, move as rapidly as it can, and work hand in glove with the other arms. On the application of these principles to the conditions which will be met with will depend the success or failure of the tank attack.

The first condition to inquire into is the position of the objective; is the ground leading up to it suitable for tank movement, is the country on the flank of such a nature as to permit of offensive wings being formed? The second is the position and number of the enemy's guns; can these be controlled by counter-battery work or smoke; how will they affect the lines of approach and their selection? The third is the number of subsidiary objectives before the final one is captured. The fourth is the "springing off" position of our own infantry, and the fifth is, how can the enemy be surprised?

These five questions being satisfactorily answered, the normal procedure is to divide the whole tank force into a main body and two wings; to take these three forces and to divide each up into as many lines of tanks as there are objectives to be attacked; to divide each objective up into tank attack areas according to the number of tactical points each contains. Provided the enemy does not possess tanks himself, or a sure antidote to their use, which the Germans never did possess, a well-considered and mounted tank, infantry, artillery, and aeroplane attack is the nearest approach to *certainty of success* that has ever been devised in the history of war. No well-planned extensive tank attack has in the past ever failed, and each one has resulted in more prisoners having been captured than casualties suffered. These are historic facts and not mere *pæans* of praise; they, consequently, deserve our most careful consideration when eventually we plan and prepare for the future.

CHAPTER 9
The Battle of Arras

The great battles which opened the Allies' 1917 campaign on the Western Front were the direct outcome of two main causes:

(1) The strategical positions of the opposing armies resulting from the Battle of the Aisne in 1914.

(2) The tactical position of the same armies resulting from the Battle of the Somme in 1916.

The former placed nine-tenths of the German Army in the west, in a huge salient Ostend-Noyon-Nancy; the latter a considerable portion of that army in a smaller one, Arras-Gommecourt-Morval. The former offered possibilities for the Allies to get in a right and left hand blow on two of the main centres of the German communications—Valenciennes and Mézières; the latter a right and left hand blow in the direction of Queant against the northern and southern flanks of the German Sixth and First Armies.

Had it been possible to bring off these latter blows successfully, such a debacle of the German forces would have resulted that not only would the advance of the British First, Third, Fourth, and Fifth Armies have seriously threatened Valenciennes, but the rush of German reserves to stop the gap would have withdrawn pressure from before the French about Reims, and would probably have enabled them to advance on Mézières.

A plan for an attack in the vicinity of Arras had been considered shortly before the opening of the Battle of the Somme on July 1, 1916; it was then dropped, only to be revived in October, when the plan contemplated was to drive in the northern flank of the Gommecourt salient. It was hoped to employ two battalions of forty-eight tanks each in this operation; but, as the tanks promised in January did not materialise until the end of April, this plan had to be continually modified.

Meanwhile a hostile operation began to take place which bid fair to filch from us the tactical advantage we had won during the preceding summer. Towards the end of February, it became apparent that the Germans intended to evacuate the Gommecourt salient; and the recent construction of the Hindenburg Line suggested a rounding off of the right angle between Arras and Craonne.

The German retirement necessitated certain changes in the British plan of operations. The Fourth Army relieved the French between the Somme and Roye; the Third Army, consisting of five Corps and three Cavalry Divisions, was now to penetrate the German defences, and by marching on Cambrai turn the Hindenburg Line from Heninel to Marcoing; the First and Fifth Armies were to operate on the left and right flanks of the Third Army.

The success of the British plan of attack depended on penetrating not only the German front-line system, but also the Drocourt-Queant line within forty-eight hours of initiating the attack; for, by so doing, so severe a wound would be inflicted that the Germans would be forced to move their reserves towards Cambrai and Douai, and away from Soissons and Reims, where the main blow was eventually to fall. Time, therefore, was, as usual, the all-important factor—could the Drocourt-Queant line be penetrated before the enemy was able to assemble his reserves?

Tanks, it was decided, should assist in gaining this time, yet on April 1, after denuding the training grounds of both England and France, only 60 Mark I and Mark II tanks could be reckoned on for the battle.

There were three ways in which these sixty tanks could be used, either by concentrating the whole against one objective such as Monchy-le-Preux, if a penetration of the centre were required, or against Bullecourt, if an envelopment of the German left flank were considered necessary, or to allot a proportion of machines to each army or corps for minor "mopping up" operations.

The last-mentioned course was eventually adopted and the following allotment of machines made:

(1) Eight tanks, to the First Army to operate against the Vimy Heights and the village of Thelus.

(2) Forty tanks to the Third Army, eight to operate with the XVIIth Corps north of the River Scarpe, and thirty-two to operate with the VIth and VIIth Corps south of the River Scarpe.

(3) Twelve tanks to operate with the Fifth Army.

The Third Army plan of operations was as follows: The VIth and VIIth Corps were to attack south of the River Scarpe between Arras and Mercatel. Their objective ran from a point 2,000 yards south-east of Henin-sur-Cojeul northwards to Guemappe, thence east of Monchy-le-Preux to the Scarpe. This objective was 10,000 yards in length and 8,000 in depth. It contained two formidable lines of defences:

(1) The Cojeul-Neuville Vitasse-Telegraph Hill-Harp-Tilloy les Mafflaines line, much of which had been fortified for over two years.

(2) The Feuchy Chapel-Feuchy line.

South of these systems was the Hindenburg Line, and east of them Monchy-le-Preux, which dominates the whole of the surrounding country. Three valleys lie between this eminence and the city of Arras.

The XVIIth Corps was to continue the attack north of the River Scarpe and occupy a line running from east of Fampoux to the Point du Jour, and thence to a point 4,000 yards east of Roclincourt. The country along the northern bank of the Scarpe was intricate, and in it many excellent positions existed for hostile machine-guns. Further, the railway running to Bailleul was in itself a formidable obstacle.

The First Army attack comprised the taking of the famous Vimy Heights, Thelus and the hill north of Thelus, a position considered one of the strongest in France.

The Fifth Army was to operate between Lagnicourt and the right of the Third Army, driving northwards towards Vis-en-Artois. The operation to be carried out by this army was a most difficult one. The destruction of the roads and the bad weather had rendered it impossible to move forward sufficient artillery—a *sine qua non* of all attacks of this period.

The whole of the above operations were to be considered as the preliminaries to the advance of two cavalry divisions and the XVIIIth Corps south of the Scarpe, which force was to break through at Monchy and advance eastwards on to the Drocourt-Queant line.

The general preparations required for a tank battle will be dealt with in another chapter, suffice it here to state that they were divided up as follows—preliminary reconnaissances, the formation of forward supply dumps, the preparation of tankodromes and places of assembly, the programme of rail movements and the fixing and preparing of the tank routes forward from the tankodromes.

Reconnaissances were started as early as January, and were most thoroughly carried out. Supply dumps were formed at Beaurains, Achicourt, near Roclincourt and Neuville St.Vaast. As no supply tanks were in existence, supplies had to be carried forward by hand and, at the time, it was reckoned that had these machines been forthcoming, each one would have saved a carrying party of from 300 to 400 men. The railheads for the Fifth, Third, and First Armies were selected at Achiet le Grand, Montenescourt, and Acq respectively. The movements of tanks and supplies to these stations were successfully carried out after several minor hitches, such as trucks giving way, trains running late, and, on March 22, 20,000 gallons of petrol being destroyed in a railway accident. Incidents such as these are, however, of little account if the plan has been worked out with foresight.

The only real mishap which occurred took place on the night of April 8–9, to a column of tanks which was moving up from Achicourt to the starting-points. Achicourt lies in a valley through which runs the Crinchon stream. The surface of the ground here is hard, but under this superficial crust lies, in places, boggy soil which was only discovered when six tanks broke through the top *strata* and floundered in a morass of mud and water. Those who were present will never forget the hours which followed this mishap. Eventually the tanks were got out, but too late to take part in the initial attack on the following day.

On April 7 and 8 the weather was fine, but, as ill-luck would have it, heavy rain fell during the early morning of the 9th. At zero hour (dawn) the tanks moved off behind the infantry, but the heavily "crumped" area on the Vimy Ridge, soaked by rain as it now was, proved too much for the tanks of the First Army, and all became

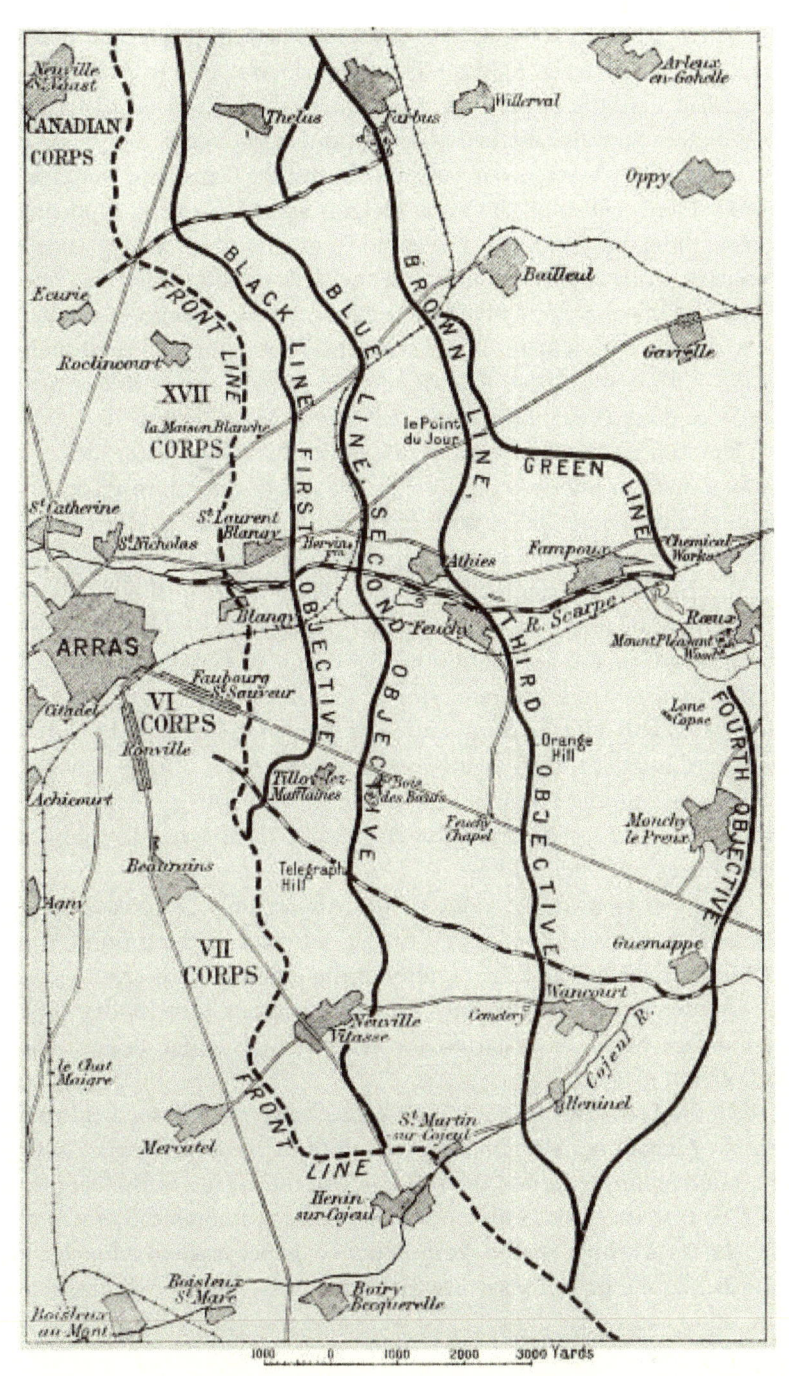

BATTLE OF ARRAS 9TH APRIL 1917.

ditched at a point 500 yards east of the German front line, and never took part in any actual fighting. The four which started from Roclincourt had but little better luck, and though they advanced considerably further they also ditched and went out of action.

The artillery barrage was magnificent and the Canadians went forward under it and took the Vimy Heights almost at a rush, capturing several thousand prisoners. The rapidity of this advance, due to the excellent work of our artillery and the dash of the Canadians, rendered the co-operation of tanks needless; it was, therefore, decided to withdraw the eight machines with the First Army, and send them to the Fifth Army. Those from Roclincourt were also withdrawn to reinforce those operating immediately north of the Scarpe.

The four tanks which started just east of Arras had better luck, for though one was knocked out by shell fire shortly after starting, the remaining three worked eastwards down the Scarpe and rendered valuable assistance to the infantry by "mopping up" hostile machine-guns.

South of the Scarpe the infantry attacked with equal *élan*. About Tilloy les Mafflaines, the Harp, and Telegraph hill the tanks caught up with the attack and accounted for a good many Germans, and then, pushing on, helped in the reduction of the Blue line (Neuville Vitasse-Bois des Bœufs-Hervin farm) and such parts of the Brown (Heninel-Feuchy Chapel-Feuchy) as they were able to reach during daylight. The ground on the Harp, an immensely strong earthwork, was much "crumped" and some of the trenches had 2 ft. of water in them. A good many tanks bellied here.

The operations of the tanks on the 9th can only be considered as partially successful—due chiefly to the difficulty of the ground, wet and heavily shelled, and the rapidity of the infantry advance.

On the following day only minor operations were undertaken, and salvage was at once started, the ditched tanks being dug out and withdrawn to refit.

On the 11th three important tank attacks were made, the first from Feuchy Chapel on Monchy; the second from Neuville Vitasse down the Hindenburg Line, and the third against the village of Bullecourt.

The first attack was eminently successful for, though only three of the six tanks which started from Feuchy Chapel reached Monchy, it was due to the gallant way in which they were fought more than to any other cause that the infantry were able to occupy this extremely valuable tactical position. Once Monchy was captured the cavalry moved forward. From all accounts the Germans, at this period of the battle, were in a high state of demoralisation, but notwithstanding this, as long as they possessed a few stout-hearted machine-gunners, an effective cavalry advance was impossible, and the only arm which could have rendered its employment feasible was the tank—the machine-gun destroyer—and as there were no longer any fit or capable of coming into action the Germans found time to stiffen their defence and to consolidate their position.

The second attack was made from Neuville Vitasse with four tanks. These machines worked right down the Hindenburg Line to Heninel, driving the Germans underground and killing great numbers of them. They then turned north-east towards Wancourt, and for several hours engaged the Germans in the vicinity of this village. All four eventually got back to our lines after having fought a single-handed action for between eight and nine hours. It was a memorable little action in spite of the fact that its ultimate value was not great.

The third operation, the attack on and east of the village of Bullecourt, is the most interesting of the three. All previous operations in this battle had been based on the timing and strength of the artillery barrage, the tanks taking a purely subordinate part. In the present attack the position of the tanks, as compared with the other arms, was reversed; for they took the leading part, and though the attack was eventually a failure, they demonstrated clearly the possibility of tanks carrying out duties which up to the present had been definitely allotted to artillery—the two chief ones being wire-cutting and the creeping barrage which, henceforth, could be carried out by wire-crushing and the mobile barrage produced by the tank 6-pounders and machine-guns.

The plan of attack was as follows: 11 tanks were to be drawn up in line at 80 yards interval from each other, and at 800 yards distance from the German line. Their task was to penetrate the Hindenburg Line east of Bullecourt; 6 to wheel westwards (4 to attack Bullecourt and 2 the Hindenburg Line north-west of Bullecourt), 3 to advance on Reincourt and Hendecourt, and 2 to move eastwards down the Hindenburg trenches. This operation was similar to the one already discussed in Chapter 7, "Tank Tactics," and called the "Trident Formation."

All 11 tanks started at zero, which was fixed at 4.30 a.m. Those on the wings were rapidly put out of action by hostile artillery fire; however, 2 out of the 3, ordered to advance on Reincourt and Hendecourt, entered these villages and the infantry following successfully occupied them.

In spite of the very heavy casualties suffered, the tanks in the centre had carried out their work successfully, when a strong converging German counter-attack, partly due to the impossibility of creating offensive flanks to our central attack, retook the villages of Reincourt and Hendecourt, captured the two tanks and several hundred men of the 4th Australian Division. The loss of the two tanks was unfortunate, for the Germans discovered that their latest armour-piercing bullets would penetrate their sides and sponsons. This discovery led to a German order being published that all infantry should in future carry a certain number of these bullets.

The interest of the Bullecourt operation lies in the fact that it was the first occasion on which tanks were used to replace artillery. It failed for various reasons—the haste with which the operation was prepared; the changes in the plan of attack on the night prior to the attack; the unavoidable lack of artillery support; and above all the insufficiency of tanks for such an operation and the lack of confidence on the part of the infantry in the tanks themselves.

Between April 12 and 22 all tank operations were of a minor nature. By the 20th of this month thirty of the original machines were refitted and on the 23rd eleven of these were employed in operations around Monchy, Gavrelle, and the Chemical Works at Rœux; excellent results were obtained, but no fewer than five out of the eleven machines sustained serious casualties from armour-piercing bullets, which had now become the backbone of the enemy's anti-tank defence.

The general result of the tank operations was favourable, though the number of casualties sustained exceeded expectation. The value of the work they accomplished was recognised by all the units with which they worked. The casualties they inflicted on the enemy were undoubtedly heavy; in most cases where they advanced the infantry attack succeeded, and the highest compliment which was paid to their efficiency came from the enemy himself, who took every possible step to counter their activity.

The operations showed that the training of all ranks had been carried out on sound and practical lines. The fighting spirit of the men was high, the tanks being fought with great gallantry. One commanding officer stated, in his report on the battle, that the behaviour of his officers and men might be summed up as "a triumph of moral over technical difficulties."

This fine fighting spirit was undoubtedly due to the excellent leadership all officers and N.C.O.s had exercised during individual and collective training; and to the full recreational training given to the battalions during these periods, games and sports as a fighting basis having been sedulously cultivated.

The main tactical lessons learnt and accentuated were:—that tanks should be used in mass, that is they should be concentrated and not dispersed; that a separate force of tanks should be allotted to each objective, and that a strong reserve should always be kept in hand; that sections and, if possible, companies should be kept intact; that the Mark I and Mark II machines were not suitable to use over wet heavily-shelled ground; that the moral effect of tanks was very great; that counter-battery work is essential to their security; and that supply and signal tanks are an absolute necessity.

On the evening of April 10, the Colonel Commanding the Heavy Branch received the following telegram from the Commander-in-Chief:

> My congratulations on the excellent work performed by the Heavy Branch of the Machine Gun Corps during yesterday's operations. Please convey to those who took part my appreciation of the gallantry and skill shown by them.

Chapter 10
Tank Battle Records

In order to record the personal experiences of each tank Crew Commander in battle, and to collect statistics as to the work of the tanks themselves, shortly before the Battle of Arras was fought, a form was introduced known as a "Tank Battle History Sheet." These sheets were issued to Crew Commanders prior to an engagement, were filled in by them after its completion and, eventually, forwarded to Tank Corps Headquarters, where they were summarised by the Tank Corps General Staff. By this means it was possible to collect many valuable experiences from the soldiers themselves, information which unfortunately so frequently is apt to evaporate when the final battle report starts on its journey from one headquarters to the next.

Outside the material value of these reports they frequently possessed a psychological value, and by reading them with a little insight it was possible to gauge, with fair accuracy, the moral of the fighting men, an "atmosphere" so difficult to breathe when in rear of the battle line, so impossible to create, and yet so necessary to the mental health of the General Staff and the Higher Command.

This system of record, initiated at the Battle of Arras, was maintained in the Tank Corps up to the conclusion of the war, many hundreds of these brief histories being written. The following are taken, almost at random, from those made out during the above-mentioned battle, and are fair examples of early tank fighting.

BATTLE HISTORY OF CREW No. D.6. TANK No. 505. DATE 9/4/17.
COMMANDED BY LIEUTENANT A——

Unit to which attached	14th Division.
Hour the tank started for action	6.20 a.m., April 9, 1917.
Hour of zero	5.30 a.m. (14th Division attacking at 7.30 a.m.).
Extent and nature of hostile shell fire	Increasing as tank worked along Hindenburg Line.
Ammunition expended	3,500 rounds S.A.A.
Casualties	Nil.
Position of tank after action	Caught in large tank trap and struck by shell fire.
Condition of tank after action	Damaged by shell fire.

Orders received.—To attack Telegraph hill with infantry of 14th Division at 7.30 a.m. on April 9, 1917, then proceed along Hindenburg Line to Neuville Vitasse. To wait at rallying-point N.E. of Neuville Vitasse until infantry advanced again towards Wancourt. To proceed with infantry to Wancourt and assist them wherever necessary.

Report of action.—Tank left starting-point at Beaurains at 6.30 a.m., on April 9, 1917, crossed our front line at 7.27 a.m., attacking Telegraph Hill with the infantry at 7.30 a.m.; then worked towards Neuville Vitasse along the Hindenburg Line. At a point about 1,000 yards N.E. of Neuville Vitasse, the tank was caught in a trap consisting of a large gun-pit carefully covered with turf. I and Sergeant B—— immediately got out and went to guide other tanks clear of the trap in spite of M.G. and shell fire.

(Signed) A——, Lieut.
O.C. Tank D.6.

BATTLE HISTORY OF CREW No. D.9. TANK No. 770. DATE 9/4/17.
COMMANDED BY 2ND LIEUTENANT C——

Unit to which attached	30th Division.
Hour the tank started for action	4.45 a.m.
Hour of zero	5.30 a.m.
Extent and nature of hostile shell fire	Very severe from the moment of entry into enemy lines.
Ammunition expended	Unknown.
Casualties	Corporal wounded, since sent to hospital.

Position of tank after action	Ditched in C.T. near Neuville Vitasse trench.
Condition of tank after action . . .	Ditched but sound.

Orders received.—To proceed from Mercatel to the Zoo trench system through the Cojeul switch to Nepal trench, from thence with the infantry to Wancourt.

Report of action.—Owing to mechanical trouble tank was delayed in coming into action. Having rectified this, I proceeded to join D.10—D.11 as ordered.

I eventually found these tanks out of action and proceeded alone to a further line of trenches, where I met with decidedly severe hostile machine-gun and shell fire. I consider we were successful in quelling one of the many sniper posts, but on account of being ditched were prevented from proceeding. It would appear, however, that the presence of my tank—it being on the right flank of our infantry, which was up in the air—was a deterrent to the enemy, of whom small bodies were still in existence in the vicinity. I caused my 6-pounders to be manned, and we held our position for three days, when the tank was eventually got out of her position. As a whole, the crew worked together well and cheerfully, but I would especially commend Corporal D—— for unfailing cheerfulness and devotion to duty under very trying and disappointing circumstances.

(Signed) C——, 2nd Lieut.,
O.C. Tank D.9.

BATTLE HISTORY OF CREW No. D.4. TANK No. 783. DATE 23/4/17.
COMMANDED BY LIEUTENANT E——

Unit to which attached	50th Division (4th Battalion Yorks. Regt.).
Hour the tank started for action . .	3.30 a.m.
Hour of zero	4.45 a.m.
Extent and nature of hostile shell fire .	Shell fire heavy, practically no shrapnel. Machine-gun fire not excessive.
Ammunition expended	Approximately 40 rounds (6-pounder).
Casualties	Nil.
Position of tank after action . . .	0.19.b.05 (approx.).
Condition of tank after action . .	Unserviceable: both tracks broken, probably other damage from direct hits; also on fire.

Orders received.—To attack enemy strong point at 0.19.a.07 as my first objective, then to proceed to banks in 0.19.b. and return with the infantry until the Blue Line was consolidated, as my second objective.

My third objective was to conform with an advance by the infantry at zero plus seven hours, and to attack a tangle of trenches in 0.21.a. & b. just in advance of the Red Line. It was eventually left to my decision as to the possibility of attempting this third objective.

Report of action.—Advanced with infantry, but owing to heavy mist had great difficulty in following exact route. Reached first objective at 5.20 a.m., having approached it from the river side. Successfully dealt with several of the enemy on left bank of river, causing them to retire. Cruised about until joined by tank No. 522, D. 3. Then proceeded towards second objective. On the way I saw our infantry retiring, went ahead to stop enemy advance. Whilst going forward I saw Lieutenant F——'s tank, which was then off its route. Lieutenant F—— came out of his tank and informed me that he had lost his way. I redirected him, and he then rejoined his tank.

Almost immediately after this (approx. 6.30 a.m.) both tanks came under direct anti-tank gun and machine-gun fire. The latter was silenced by my left 6-pounder gun. I manoeuvred to present as small a target as possible to the former. The tank, however, received about six direct hits, which damaged both tracks, set alight the spare petrol carried in box in rear of tank, and possibly caused other serious damage. The whole crew succeeded in escaping from the tank unhurt. Position of tank as stated.

I then returned to Coy. H.Q. and reported.

(Signed) E——, Lieut.,
O.C. Tank D.4.

BATTLE HISTORY OF CREW No. D.10. TANK No. 784. DATE 23/4/17.
COMMANDED BY 2ND LIEUTENANT G——

Unit to which attached	98th Infantry Brigade.
Hour the tank started for action	4.45 a.m.
Hour of zero	4.45 a.m.
Extent and nature of hostile shell fire	First three hours artillery fire not very heavy, but from then very heavy fire until rallying-point was reached. No direct fire by anti-tank guns.
Ammunition expended	290 rounds 6-pounder, remainder on tank could not be used owing to the shells sticking in shell casings on tank. Eight pans for Lewis-gun ammunition.
Casualties	Nil.
Position of tank after action	Factory Croisilles, 12 noon.
Condition of tank after action	Good—only required refilling and greasing.

Orders received.—To advance from starting-point on British front line at T.4.b.4.5 to Hindenburg Line at point T.6.a.0.5, from which point infantry were to bomb along Hindenburg Line (front and support) to River Sensée at U.7.a.4.4. Tank to assist infantry and after objective at river taken to proceed to Croisilles.

Report of action.—I started from starting-point at T.4.b.4.5 at zero, and made for Hindenburg wire at T.6.a.0.5, crossing same and getting into touch with our infantry, from whom I received report that they were held up by machine-guns along the trench. I proceeded to this point and cleared the obstacle. I then travelled parallel to the trench, knocking out machine-gun emplacements and snipers' posts all the way down to point U.1.c.5.0. The infantry kept in touch all the way down, moving slightly in rear of tank, and after emplacements were knocked out, they took the occupants prisoners. In two cases white flags were hoisted as soon as the emplacement was hit. The shooting was very good. Up to point U.1.c.5.0 the shelling had been casual, but when we reached the N. bank of the sunken road at this point and were firing into emplacements towards the river we were in full observation from the village and the artillery fire became very heavy.

The supply of 6-pounder ammunition now became exhausted, and the ground on the S. side of sunken road being very bad, I decided to move back along the trench and then crossed the wire, and crossing sunken road at about T.12.b.5.3, made for rallying-point at Factory at Croisilles, where I arrived at 12 noon. I was shelled heavily all the way back to the rallying-point, but no damage was done. I was of opinion that the Hindenburg front line was too bad (wide) to cross, and so could not deal with support line and was unable to observe this line from front line. I sent two pigeon messages at 9.30 a.m. and 12 noon. I had only one message clip, so had to fasten second message with cotton.

(Signed) G——, 2nd Lieut.,
O.C. Tank D. 10.

BATTLE HISTORY OF CREW NO. D.9. TANK NO. 770. DATE 9/4/17.
COMMANDED BY 2ND LIEUTENANT C——

Unit to which attached	30th Division.
Hour the tank started for action	4.45 a.m.
Hour of zero	5.30 a.m.
Extent and nature of hostile shell fire	Very severe from the moment of entry into enemy lines.
Ammunition expended	Unknown.

Casualties	Corporal wounded, since sent to hospital.
Position of tank after action . . .	Ditched in C.T. near Neuville Vitasse trench.
Condition of tank after action . .	Ditched but sound.

Orders received.—To clear Mount Pleasant wood, Rœux, and northern edge of village.

Report of action.—Time allowed for tanks from deployment point to starting-point proved to be insufficient, which delayed my start some twenty minutes. Having learnt that the other car which was operating with me was "out of action," I made my way alone to the railway arch, where I was held up some few minutes owing to a number of stretcher cases which had to be removed, and a sand-bag barricade which I could not push down.

I soon caught up the infantry, who were held up by machine-gun fire in Mount Pleasant wood. At their request I altered my course and made for the northern side of the wood running parallel with the trench which we held at the south of the wood, and which the enemy held at the north. I was told that a bombing party would follow me up the trench.

Having cleared this wood, I pushed on towards the village of Rœux, where I again met the infantry who had come round the other side of Mount Pleasant wood, where they were again held up by machine-gun fire which came from the buildings.

Our barrage could only have been very slight, to judge from the comparatively small amount of damage which was done to the buildings. Here I used 200 rounds of 6-pounder ammunition.

It is difficult to estimate with any accuracy the number of machine-guns actually "put out." One of my best targets was a party of some thirty men whom we drove out of a house with 6-pounders, and then sprayed with Lewis-gun fire.

I am sure that at least one 6-pounder shell dropped amongst these—this made a distinct impression.

Another target that presented itself was a party of men coming towards us. I do not know whether they intended giving themselves up or whether they were a bombing party—I took them for the latter.

Parties were frequently seen coming up from the rear, through gaps in the buildings.

Twice an enemy officer rallied some dozen or so men and rushed a house that we had already cleared. Here again a 6-pounder through the window disposed of any of the enemy remaining in the buildings.

In regard to the machine-guns in the wood, we could only locate them by little puffs of smoke at which we fired our 6-pounders. We did not take our departure until these puffs had disappeared, and there was in consequence reasonable ground to suppose that the guns had been knocked out. Finally, our infantry reached the village. Apparently, there was no officer commanding our infantry in this part of the line.

I then moved towards Rœux wood and learnt of a sniper still left in Mount Pleasant wood and a machine-gun, which was causing great trouble, on the railway embankment, and I then made my way back to the railway arch with a view to running parallel with the embankment towards the station, but unfortunately my car bellied in the very marshy ground by the canal. With regard to casualties, it is my opinion that I was in the district sufficiently long enough, some three hours, to enable the enemy to send for a supply of armour-piercing bullets. All four of my crew were hit whilst in the car.

The Lewis-gun mountings were bad, many targets were lost owing to the time it took to mount the gun, and finally we mounted the gun through the front flaps. The flap of the present mounting does not rise high enough to clear the foresight.

Both the 6-pounder guns worked splendidly, only giving one misfire the whole time. There was no hostile shelling of any kind in the village of Rœux or immediately in the district where I was operating, but the enemy barrage falling round the railway was of a very severe nature. When I found that it was impossible for me to proceed towards the railway station, I sent off a pigeon message at the request of an O.C. requesting that one of the cars operating in the Chemical Works district should be detailed to deal with the machine-gun on the embankment.

I cannot speak too highly of the efficiency and general work of the crew.

I have handed two German diaries, which came into my possession, to the Company Intelligence Officer.

(Signed) H——, 2nd Lieut.,
O.C. Tank No. 716.

Chapter 11
The Second Battle of Gaza

On account of the assistance rendered to the British infantry by tanks during the Battle of the Somme a decision was arrived at in

England to despatch a number of these machines to Egypt to assist our troops in the Sinai Peninsula, especially in the neighbourhood of El Arish, south of the Turkish frontier. The number originally decided on was twelve, but this was eventually cut down to eight, and, through an unfortunate error, old experimental machines were sent out instead of new ones as intended.

The detachment, under the command of Major N. Nutt, consisted of 22 officers and 226 other ranks drawn from the original E Company, and together with its tanks, workshops, and transport, it embarked at Devonport and Avonmouth in December 1916, arriving in Egypt during the following month.

Demonstrations and schemes were at once arranged for so that the staffs of the various fighting formations could witness what tanks were able to accomplish. These schemes were carried out on the sand dunes near Gilban, some ten miles north of Kantara on the Suez Canal.

In February, orders were suddenly received one day for the detachment to move with all possible speed to the fighting zone.

******

Major O. A. Forsyth-Major (Second in Command of the Egyptian Tank Detachment), on whose report this chapter is based, lost all his documents and maps at sea in May 1918 when the ship on which he was returning to England was torpedoed and sunk, consequently some of the dates are missing.

******

This was carried out, and within three hours of receiving the orders the entire detachment, with tanks and accessories, had entrained at Gilban, and was speeding northwards towards the area of operations. Next day a delay occurred at El Arish, which the day previously had been captured by the Australians; but, the same evening, the train proceeded to Rafa, a frontier town, which had only just been evacuated by the Turks, and early next morning reached Khan Yunus, some fifteen miles south-west of Gaza, an old Crusader stronghold surrounded by vast fig groves and other vegetation; here the detachment remained for ten days.

During this halt the First Battle of Gaza had come to an end, our troops having been obliged to retire and take up a position to the south of the town owing to the appearance of strong Turkish reinforcements from the direction of Beersheba; these threatened the British communications.

Hostilities now ceased and preparations were begun for the Second Battle of Gaza, which was to prove one of the fiercest contests of the war in its eastern theatre. For this battle, early in March, the Tank Detachment moved from Khan Yunus to Deir el Belah.

The Turkish Army at this period, numbering some 30,000 men, was disposed along a sixteen-mile front extending from Gaza south-eastwards to Hareira and Shekia. The British plan of operations was as follows:

The G.O.C. Desert Column was entrusted with the operations against the Hareira front, protecting the right flank, whilst the task of seizing the important ridges of Sheikh Abbas and Mansara, both commanding Gaza and situated to the south of this town, was assigned to the 52nd, 53rd, and 54th Divisions; the 74th Division remaining in general reserve.

The tanks of the detachment, which had been held in G.H.Q. reserve, were now allotted to divisions as follows:

(1) 53rd Division, operating from the sea to the Cairo road, running through Romani trench: two tanks which were to be held in reserve until the infantry had advanced to the line—Red House-Tel El Ajjul-Money House-the coast.

(2) 52nd Division, operating from Kurd valley to Wadi El Nukhabir: four tanks to support the infantry attack on the Mansara ridge.

(3) 54th Division, operating on a front extending from 500 yards west of Abbas ridge to the Gaza-Beersheba road: two tanks to support the infantry attack on the Sheikh Abbas ridge.

Z day was to be April 17. Two days prior to this the eight tanks left Deir El Belah after dusk, two proceeding over the Druid ridge through St. James's Park, thence by Tel El Nujeid across the Wadi Ghuzze to Money hill; four from Deir El Belah in an easterly direction through Piccadilly Circus over the prominent ridge of In Seirat, then eastwards to Sheikh Nebhan on the Wadi Ghuzze; two followed the same route as far as In Seirat, and from there made for a point south-east of Sheikh Nebhan.

All eight tanks reached their positions of assembly before dawn without mishap and in good condition. Meanwhile ammunition and supply dumps had been established at various spots close to the Wadi Ghuzze.

In the battle which now ensued the position of the tanks in relation to the infantry varied according to the nature of the ground and the resistance of the enemy. The attacks of the 53rd and 52nd Divi-

THE SECOND & THIRD BATTLES OF GAZA
17th April 1917 & 1st November 1917.

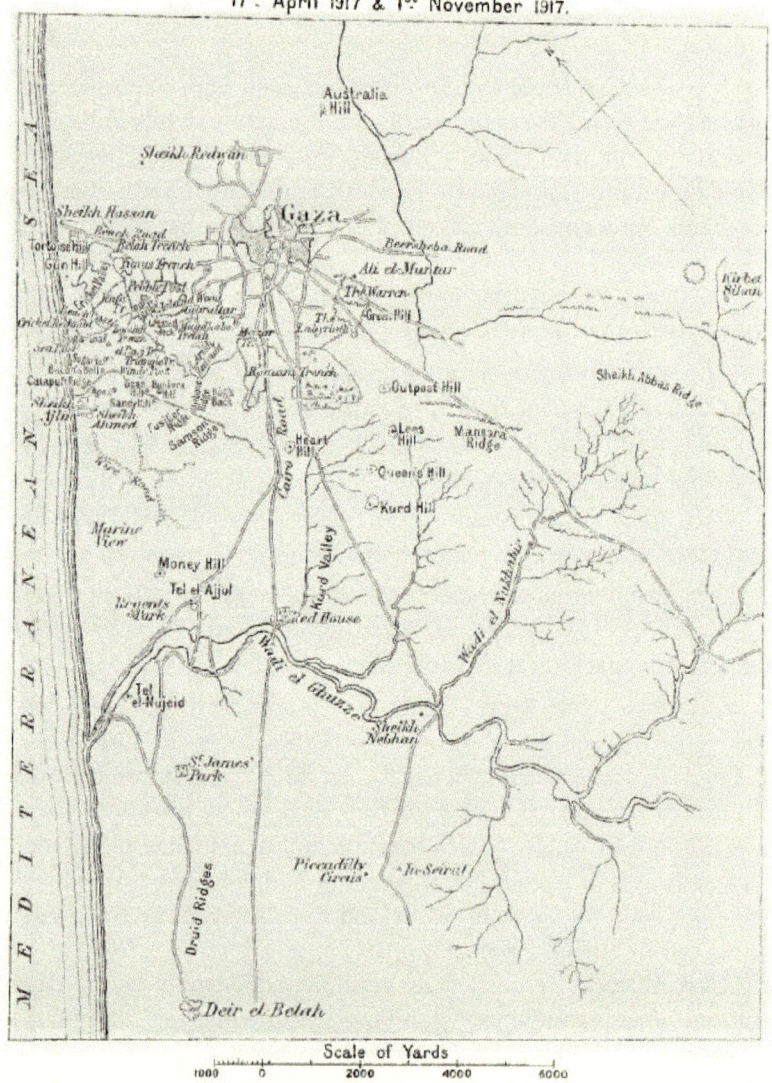

sions came as a complete surprise, the two tanks allotted to the former moved to a position south of Money hill on the evening of the first day, and the four with the latter reached a point south of the Mansara ridge. None of these machines came into action as the Turks retired from their trenches and strongholds in complete confusion. On the 54th Division's front both of the tanks allotted to this Division came into action; one, however, received a direct hit and was destroyed, but the other did good work in clearing the enemy's trenches north-west of the Abbas ridge, killing many Turks and enabling our infantry to occupy these defences.

On the evening of April 17, the three attacking divisions entrenched themselves on the line running approximately from Marine View, on the coast, through Heart Hill-Kurd Hill-Mansara-Abbas, and thence south-east to Atawineh ridge. A pause of forty-eight hours now took place wherein to prepare for the second phase of the battle.

On the morning of April 19, this second phase opened. The Australian Corps, on the right flank, was to deliver an attack on the eastern defences of Gaza, whilst the 52nd, 53rd, and 54th Divisions were to constitute the main attack, and to advance on a line running from the coast to the stronghold of Ali El Muntar. Battleships were to co-operate in this attack.

Tanks were allotted to divisions as follows:

(1) 53rd Division, objective—Mazar trench to Sheikh Redwam; one tank to assist in the capture of Sampson ridge, El Arish and Sheikh Redwam redoubts, and one tank to operate against Sheikh Ajlin, Belah-Yunus-Rafa-Zowaiid-El Burs trenches and to await further orders at El Arish trench.

(2) 52nd Division, objective—the enemy's trenches from Queen's hill to Ali El Muntar. For this operation four tanks were allotted, and their objectives, which were Outpost hill, the Labyrinth, the Warren and Ali El Muntar, were changed during the night of the 18th–19th. This resulted in considerable confusion. According to the change one tank was to precede the assault on Green hill, one to clear Lees hill and Outpost hill, and the remaining two to be kept in reserve at Kurd hill.

(3) 54th Division, objective—Kirbet El Sihan and El Sire-Ali El Muntar ridge as far as Australia hill; one tank to seize the redoubt west of Kirbet El Sihan.

From the above it will be seen a good deal was expected of the tanks, in fact these seven machines were to tackle a problem which in France would have been considered distinctly formidable for two complete battalions.

Of the two tanks with the 53rd Division one broke its track, consequently the other—the Tiger—led the advance alone and drove the enemy from Sampson ridge, which was then occupied by our infantry; it then proceeded to El Arish redoubt, but, the infantry being unable to follow, after six hours' action, during which it fired 27,000 rounds of S.A.A., it withdrew to Regent's Park, all its crew having been wounded. On the front of the 52nd Division a desperate battle took place: the tank operating against Lees hill and Outpost hill fell into a gully, the sides of which unexpectedly collapsed. Its place was taken by the tank detailed for Green hill; Outpost hill was reached and cleared, when this machine received a direct hit.

The enemy's machine-gun fire was now intense, so one of the reserve tanks was ordered up. After desperate losses the infantry eventually captured the hill, only to be driven off it by a counter-attack; they then withdrew to a line passing east and west through Queen's hill, the reserve tank withdrawing at the same time to Kurd hill. In the attack delivered by the remaining division, the 54th, no better luck was experienced. The one machine working with this division moved on the great redoubt north-west of Kirbet El Sihan, and reaching this work the Turkish garrison surrendered. The infantry then took over the position. Shortly after this a direct hit broke one of the tracks of this tank, and a counter-attack eventually resulted in its capture with the infantry who had occupied the redoubt.

In spite of the fact that this battle was unsuccessful, the work carried out by the Tank Detachment constitutes a remarkable feat of arms. The tanks engaged were Mark I's and II's, which, by the time the battle was ended, had each covered on an average some 40 miles of country. Reconnaissance, due to want of time, was practically non-existent, and the limitations of the tank were not understood by the infantry commanders, who expected miracles from a far from perfect machine. The objectives allotted were not only difficult, but too numerous, yet in spite of this the protection which these eight tanks afforded the attacking infantry on a five-mile frontage was considerable and fully appreciated; it was, however, quite inadequate on account of the hundreds of ingeniously hidden machine-guns, to which the Turks mainly owed their victory.

Chapter 12
Staff Work and Battle Preparation

The foundations of the success or non-success of a battle rest on its organisation, that is, on the preparations made for it. This is the duty of the General and Administrative Staffs of an Army or Formation and usually entails an immense amount of careful work. The fact that success depends as much, if not more, on organisation (brain power) as on valour (nerve power) is not generally recognised, and many an officer and man in the firing line is, through ignorance of the causes and effects which are operating behind, only too prone to forget what the staff is doing, and, never more so, what the staff has done than after a really great victory has been gained.

The more scientific weapons become the more will good staff work decide whether their use is going to lead to victory or defeat. This was very early realised in the Tank Corps, and every endeavour was made by its commander and his subordinate leaders to select only the most capable officers for their respective staffs; this resulted in ability more often than seniority deciding the filling of an appointment.

The work of the Staff of the Tank Corps was often considerably complicated by the fact that, the tank being a novel weapon of war, it was little understood, not only by the other arms, but by many members of the Tank Corps itself. This resulted in a great deal of educational work being required before many measures, very obvious to the Staff itself, were accepted by others. In the early days of the Corps the tank was generally placed by the other arms under one of two categories—a miracle or a joke, and this did not tend to facilitate or expedite preparations.

The main duty of the General Staff is to foresee by thinking ahead, of the Administrative Staff to prepare, of the Commander to decide, and of his troops to act. These four links go to build up a battle, and if any one of them is defective, the whole chain is weak. On the power of thinking ahead, that is, foreseeing conditions and events, will depend all preparations. Decisions cannot with safety be simply based on former experiences, codified and printed rules and regulations; if this were possible every intelligent subaltern with a good memory or a big pocket could become a Napoleon in six weeks. ("It is not some familiar spirit which suddenly and secretly discloses to me what I have to say or do in a case unexpected by others; it is reflexion, meditation."—Napoleon.)

Decisions must be based on weapons and men moved in accordance with the principles of war as governed by conditions existing and possible. Possible conditions cannot be guessed at; if they could the planchet board and not the baton would be the emblem of a field-marshal's worth. Possible conditions can only be guarded against or converted into allies by being prepared to meet all eventualities, and these preparations in a formation such as the Tank Corps were at first prodigious. Little by little, however, the knack of mechanical warfare was cultivated, and then what had at first appeared a mountain eventually turned out to be a molehill—a good sprinkling of molehills, and not a few mountains, however, remained over even to the last.

In the training schemes carried out before the Battle of Arras, as mentioned in Chapter 7, no fewer than 132 headings of various measures, all relative to the preparations requisite for an offensive with tanks, were laid down. After this battle this number was considerably increased and continued to grow as each engagement added new experiences to the old ones.

The main preparations required for an offensive are the following: (1) Movements, (2) Reconnaissances, (3) Secrecy, (4) Supply, (5) Communication, (6) Assembly, (7) Tactics, (8) Reorganisation.

Tank movements generally fall under the headings of rail movements and cross-country movements. As regards the former it must be remembered that the tank cannot at present move over lengthy distances under its own power. A Mark I tank on good going could not be relied on to run more than 12 miles on a fill of petrol, and after it had completed about 70 miles it had to be overhauled and many of its parts renewed. Rail movements require special trucks, and, in the early days of the Corps, special sidings and entraining ramps. As regards the latter, cross-country routes should be reconnoitred beforehand. It will be remembered how on the night of April 8–9, the night previous to the first day of the Battle of Arras, six tanks became ditched at Achicourt on account of a bog existing under the hard surface of the ground. Had the officer reconnoitring this route tested the ground along the valley by pushing a stick into it, this accident would not have occurred, for the stick would have penetrated the crust and informed him of the nature of the soil below it.

Before any move takes place from the position of assembly, near railhead, to the starting-points, the points whence the tanks will proceed into battle, the following are a few of the subjects that a Tank Unit Commander will have to consider:—Objectives; strong-points;

machine-gun emplacements; batteries; trenches; wire; infantry lines of advance; minimum number of tanks required for the main objectives; minimum of tanks required for subsidiary objectives; nature of ground and its probable condition at zero hour; where ground, soil, natural features and hostile batteries will chiefly impede tank movements; the lines of least resistance for tanks through the enemy's lines to the main objectives and the points of greatest resistance to the infantry advance; landmarks; starting points with reference to lines of least resistance; positions of deployment with reference to starting-points; tank routes from positions of assembly to the positions of deployment and thence to the starting-points; any places on these tank routes where delays are likely to occur; rallying points; supply dumps; communication, etc., etc.

So in turn must each move or preparatory measure be dealt with, reconnaissance playing an all-important part, not only before the battle, but during it, and immediately after it, and if the system of communication during the battle is not efficient the work of the reconnaissance officer will frequently be wasted, so we find one preparation depending for its worth on another until the whole forms a complete and somewhat intricate chain.

Imagine now, when this chain is nearing completion what it means to it if some new plan be evolved, or a change be introduced or forced on to a scheme of operations—its effect will frequently have to be carried right down the chain, and this will not only mean new work being done, but old work being undone. Take the following as an example: a battalion of tanks is to detrain at A, a few days later it is ordered to detrain at B instead; this will probably entail shifting 20,000 gallons of petrol, 12,000 6-pounder shells, 300,000 rounds of S.A.A., and countless other stores. It is these changes in operations which a good Staff guards its troops against by foresight; this being so, the efficiency of a staff may usually be gauged by the number of amendments a commander issues to his orders.

Another important duty of the staff is to assist the troops when the period of preparation ends and action begins, and further still to watch closely every action so that changes may be foreseen and preparations may be improved on the next occasion. These duties are called "Battle Liaison," a duty which was impressed upon every General Staff Officer in the Tank Corps as the most important he would be called upon to carry out. At every battle from that of the Ancre onwards, the majority of the headquarters' General Staff Officers were present on the battlefield itself, not after the fight had swept on but before it opened and whilst it lasted. Each night these officers would report to their headquarters not only what they had heard but what they had seen, a much more reliable source of evidence.

The result of this system was that when the crash came on March 21, 1918, though the Tank Corps was split up over a front of 60 miles, and in many places complete confusion followed the German attack, not once from that day on to the end of the battle did the headquarters of the Tank Corps lack information regarding the position of all its units. This may be chronicled as a notable "feat of staff work," and certainly as useful as many a much more spectacular "feat of arms." It is for this type of staff work that Staff Officers are sometimes rewarded.

Chapter 13
The Battle of Messines

The situation at the end of April 1917 was a difficult one for the Allies. The failure to penetrate the Drocourt-Queant line had rendered the whole plan of the British attack east of Arras abortive; this was bad enough, but indeed a minor incident when compared with the failure of the great French attack in Champagne. It was towards making good this failure that the rest of the year's operations had to be directed.

The ambitious plan of cutting off the Arras-Soissons-Reims salient having failed, the next blow was to be directed against the German right flank. The object of this attack was to drive this flank back sufficiently far to deprive the Germans of the coast line between Nieuport and the Dutch frontier, and to render their position about Lille sufficiently insecure to force them to evacuate it and so open the road to Antwerp and Brussels.

The possibility of such an operation as this had long been contemplated, and, as early as the summer of 1916, preparations for it had been taken in hand by the British Second Army. By May 1917 these were completed, including the construction of an extensive system of railways in the Ypres area and the mining of most of the western flank of the Messines-Wytschaete ridge.

The operation was to be divided into two main phases, firstly the taking of the Messines-Wytschaete ridge so as to secure the right flank of the second phase and to deny the enemy important points of observation, and secondly the attack north of this ridge with the object of occupying the coast line and of pushing forward towards Ghent. The first phase constituted the Battle of Messines, the subject of this chapter, and the second the Third Battle of Ypres.

It had been foreseen early in the year that such an attack was possible, and that in all probability tanks would be called upon to take part in it, consequently, as early as March 1917, reconnaissances of the whole of the Ypres area had been taken in hand by the Heavy Branch. In April this surmise proved correct, and the 2nd Brigade, consisting of A and B Battalions, was selected for this operation. In May these two battalions were equipped with thirty-six Mark IV tanks each.

Railheads were selected at Ouderdom and at Clapham Junction (one mile south of Dranoutre), and though they were within the shelled area, they fulfilled most of the requirements demanded of a tank railhead. Advanced parties began arriving at these stations on May 14, and between May 23 and 27, A and B Battalions followed them. Supply dumps were then formed, and arrangements were made to carry forward one complete fill for all tanks operating by means of Supply Tanks, which were first used in this battle. These tanks consisted of discarded Mark I machines with specially made supply sponsons fitted to them.

The positions of assembly were selected quite close to railheads, B Battalion tanks being hidden away in a wood, and A Battalion's in specially built shelters representing huts. The spoor left by the tanks, as they moved to these positions, was obliterated by means of harrows so that no enemy's aeroplane happening to cross over our lines this way would notice anything suspicious on the ground.

The object of the Second Army's operations was firstly to capture the Messines-Wytschaete ridge and secondly to capture the Oosttaverne line, a line of trenches running north and south a mile to a mile and a half east of the ridge. Three Corps were to participate in this attack—the Xth and IXth Corps, and the IInd Anzac Corps. To these Corps tanks were allotted as follows: 12 tanks to the Xth Corps, 28 tanks to the IXth Corps, 16 for the ridge, and 12 for the Oosttaverne line, and 32 tanks to the IInd Anzac Corps, 20 for the ridge and 12 for the Oosttaverne line. Each battalion had two spare tanks and 6 supply tanks. The total number of tanks used was, therefore, 76 Mark IV tanks and 12 Mark I and II Supply Tanks.

The tank operations were planned to be entirely subsidiary to the infantry attack, in fact the whole attack, being limited to a very short advance, was based on the power of our artillery and the moral effect produced by exploding some twenty mines simultaneously at zero hour.

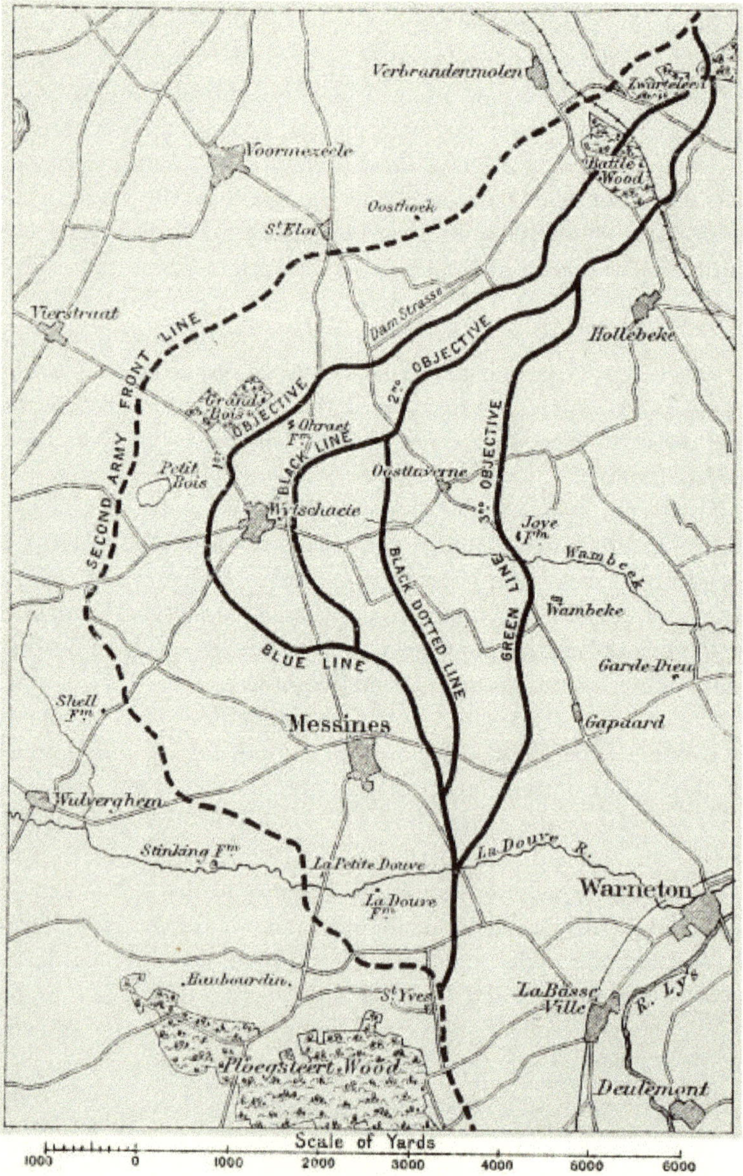

Verbrandenmolen
Zwarteleen
Voormezeele
Battle Wood
Oosthoek
St Eloi
Dam Strasse
Vierstraat
Hollebeke
SECOND ARMY FRONT LINE
1st OBJECTIVE
2nd OBJECTIVE
3rd OBJECTIVE
Grand Bois
Obrast Fm
BLACK LINE
Oosttaverne
Petit Bois
Wytschaete
Joye Fm
Wambeek
BLACK DOTTED LINE
GREEN LINE
Wambeke
BLUE LINE
Garde Dieu
Shell Fm
Messines
Gapaard
Wulverghem
Stinking Fm
La Douve R.
La Petite Douve
Warneton
La Douve Fm
Hamboardin
St Yves
La Basse Ville
R. Lys
Ploegsteert Wood
Deulemont
Scale of Yards
1000
0
1000
2000
3000
4000
5000
6000

For three weeks the weather had been dry and fine, and had this not been the case, there would have been little hope of ever being able to move the tanks forward over the pulverised ground. The artillery bombardment opened on May 28; June 7 being fixed for the attack. It was a terrific cannonade, watching it from Kemmel Hill or the Scherpenberg, today almost blasted out of recognition themselves, one could see the *Grand Bois*, Wytschaete Wood, and the green fields along the valleys of the Steenbeek and Wytschaetebeek being slowly converted into a dun-coloured area which the first heavy fall of rain would convert into a porridge of mud. Some shells as they exploded would throw up great fan-shaped masses of debris and smoke, others would burst into vortex rings, whilst others again shot up into the air great feathers of fine brown dust. Day and night the bombardment continued except for a short pause now and then to mislead the enemy as to the hour the infantry would "top the parapet."

At zero hour (dawn) 40 tanks were launched; of these 27 reached the first infantry objective, known as the Blue Line, and, of these 27, 26 went on to the second objective, the Black Line, and 25 reached it.

The artillery bombardment and creeping barrage proved so effective that few of these tanks were ever called upon to come into action except round the ruins of Wytschaete village, where some snipers and machine-guns were silenced by them, and at Fanny's farm, near Messines, where our infantry were held up by machine-gun fire. One tank operating with the Anzac Corps got across the enemy's trenches at a very rapid rate and reached its objective on the Black Line, a distance of about 3,000 yards, in 1 hour and 40 minutes, having engaged an enemy's machine-gun on the way. Another tank, rather aptly named the "Wytschaete Express," led the infantry into the village of Wytschaete and helped to persuade the Germans defending it to surrender, which they did in large numbers.

At 10.30 a.m. the 24 reserve tanks were moved up to points behind our original front line, and 22 of these started with the infantry at 3.10 p.m. in the attack on the Oosttaverne line, which constituted the final objective. During this phase of the battle the tanks rendered very great assistance to the infantry by occupying the ground beyond the Oosttaverne line before the infantry arrived, and so disorganising the enemy's defence.

At a place named Joye farm an interesting incident occurred. Two tanks became ditched here, but, in spite of this and approaching darkness, these machines were in a position to repel any hostile counter-attack coming up the Wambeke valley, so it was decided to stand by in the tanks and resume unditching work at daybreak.

At about 4 a.m. unditching was started again, and one tank was got out successfully, but unfortunately broke its track soon afterwards in trying to cross the railway to advance against an enemy's counter-attack which was developing. An hour later the enemy was observed to be massing in the Wambeke valley. The position of the tanks only allowed of a pair of 6-pounders being trained in the direction of the enemy, so the remainder of the crews, under their officers, took up positions in shell-holes with their Lewis guns. Word was sent to the infantry to warn them and ask for co-operation, and on a reply being received that they were short of Lewis-gun ammunition, some was supplied to them from the tanks.

From 6.30 a.m. onwards the enemy made repeated attempts to advance, shooting at the tanks with a large number of armour-piercing bullets which failed to penetrate. They were driven off in every case with heavy casualties, until at 11.30 a.m. our artillery barrage opened and dispersed them.

The Battle of Messines, one of the shortest and best mounted limited operations of the war, was in no sense a tank battle. Tanks took but a small part in it, except in its final stages, nevertheless many useful lessons were learnt, the chief of which were: the necessity of some special form of unditching gear; the advisability of selecting starting-points well behind our front line (in this battle they were chosen too close behind it, with the result that as it was still dark at zero hour several tanks got ditched in "No Man's Land"); the advisability of selecting rallying-points well behind objectives, so that when tanks have finished their work their crews may gain as much rest as possible.

CHAPTER 14
A Tactical Appreciation

The Battle of Messines may be looked upon as the high-water mark of the artillery attack, which was first developed by the British Army during the Battle of the Somme. The time, however, was approaching when a change of tactics became imperative on account of the enemy having learnt his lesson. To appreciate what this question involves is of some interest, especially so, as in the tank was eventually discovered a means of overcoming the counter-measures now adopted by the enemy. (This chapter is extracted from a project submitted by Headquarters Tank Corps on June 11, 1917. It correctly visualised the Third Battle of Ypres, and the German artillery tactics adopted during it.)

The main characteristic which differentiates the German defensive tactics of 1917 from those of 1916 would appear to lie in the grouping of their men rather than in the siting of their trenches.

In 1916 the major portion of the German Army was placed in the frontal defensive belt because security was sought for in the maintenance of an unbroken front. In 1917, however, this security was more economically guaranteed by holding behind this front, instead of in it, a large reserve which could strike at any opponent who broke through.

This reversion to the "big idea" and the abandonment of the smaller one, namely, that war is a "series of local emergency measures," placed a further difficulty in the way of the attacker. In 1917 it was no longer a question of breaking through a defensive line as in 1914, or a zone of defences as in 1915 and 1916, but of exhausting the enemy's reserves before undertaking either of these operations with decisive effect.

This could now only be accomplished by hitting the enemy at a point which he must hold on to because of its importance or of surprising him at points where he did not expect to be attacked. If such points were not selected all he need do was to fall back as he had already done in March, and so dislocate our operations, by temporarily denying us the use of our guns.

As hitherto, the change we have always most carefully to watch for is any change the enemy is likely to carry out in his artillery tactics, and the following is apparently what the German was now doing.

Having learnt in 1916 and the first half of 1917 that if the attacker makes up his mind to do it, he can carry, by means of artillery and infantry alone, several lines of trenches in one bound, it stood to reason that the German General Staff would not continue to jeopardise its artillery by so placing it that it could be pounded to pieces during our preliminary bombardment.

If now the Germans withdrew their guns further back to a position from which, though they cannot cover their front-line system, they can cover their second or third lines and simultaneously be immune, or to a great extent immune from our counter-battery fire, by accepting the loss of a small belt of ground they would place our attacking infantry in such a position that whilst it feels the full effect of their artillery, it is receiving next to no protection from its own.

The construction of their defensive systems in 1917 did not altogether lend itself to these tactics, the systems were too close together;

but should these distances be enlarged the disadvantage to the attacker becomes apparent; and there were already signs that the Germans were fully aware of the advantage of this enlargement. At Arras they had been surprised in spite of the lengthy bombardment and they lost over 200 guns, at Messines they lost 67, and later on at Ypres only 25, on the first day of each attack. They were, in fact, countering by gun fire the exploitation of a penetration. This system of tactics can be graphically illustrated as follows (see diagram 16).

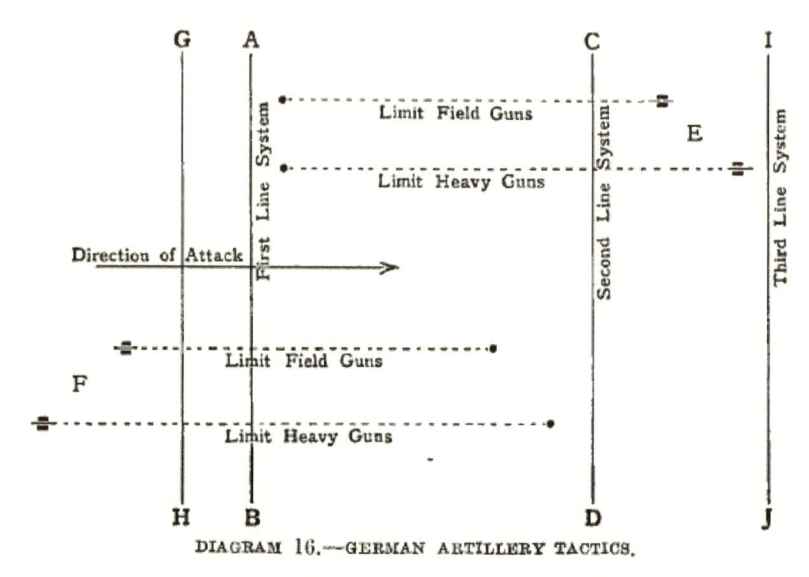

DIAGRAM 16.—GERMAN ARTILLERY TACTICS.

Suppose that AB be the German front-line system, and that CD, their second line, be so placed that the German guns at E can heavily shell the whole of CD, and yet, on account of the distance away, remain practically immune from our guns at F. Suppose also that the area ABCD is strongly wired and well sprinkled with machine guns, who is going to suffer most—the attackers from GH, who will not only be perpetually worried by the machine guns and sharpshooters in ABCD, but who will come more and more under the enemy's gun fire as they proceed towards CD, or the enemy's machine gunners occupying ABCD, and his infantry in dug-outs along CD? Undoubtedly the former, for they present the largest target, and against them is being thrown the greater number of projectiles. Suppose now the attackers capture CD, then at best they will only be able to remain there as impassive spectators to their own destruction until such time as the guns at F move forward, which, on account of the "crumped area," ABCD, will take many days. This is probably what would have happened had the Second Army been required to push the attack from the Oosttaverne line towards Wervicq.

Except for an attack on an objective of very limited depth the artillery attack was almost doomed to fail on account of slowness of moving forward the guns, due to the destruction of roads and terrain during the initial bombardment. Bearing the above possibilities in mind, it was with some apprehension that the Heavy Branch watched the approach of the next great battle, for though the "crumped" area could, if dry, be crossed by tanks, whatever service they might afford would be rendered useless on account of the impossibility of keeping the attacking infantry supplied; to do so, fleets of supply tanks would be required, and these did not exist.

CHAPTER 15
The Third Battle of Ypres

Towards the middle of May it was decided that all three Brigades of Tanks, that is, the whole Heavy Branch, should take part in the forthcoming operations of the Fifth Army east of Ypres, and that two of these brigades should assemble in Oosthoek wood and the third at Ouderdom. To initiate preparations an advanced headquarters was opened at Poperinghe early in June, and on the 22nd of this month the brigade advanced parties moved to the Ypres area.

After the Battle of Messines, the 2nd Brigade (A and B Battalions) had assembled at Ouderdom, so the present concentration only involved moving the 1st Brigade (D and G Battalions), and the 3rd Brigade (F and C Battalions) to Oosthoek. Seven trains were required for each of these brigades.

At about this time it was decided that the 1st Brigade should be allotted to the XVIIIth Corps, the 3rd Brigade to the XIXth Corps, and the 2nd Brigade to the IInd Corps; brigade commanders were, therefore, instructed to place themselves in touch with these corps and commence preparations.

The preparations required varied considerably from those for former battles. The Ypres-Comines canal, running parallel to the front of attack, formed a considerable obstacle to tank movement; consequently, causeways had to be built over the canal as well as over the Kemmelbeek and the Lombartbeek. This work was carried out by the 184th Tunnelling Company, which was attached to the Heavy Branch for the purpose. The work this unit carried out, normally under severe shell fire, was most efficient and praiseworthy. Besides the building of these causeways and the usual supply preparations, thoroughly efficient signalling communication was arranged for, including the use of a certain number of tanks fitted with wireless installations.

The reconnaissance work of the battalions was greatly facilitated by that already carried out by the advanced headquarters party. Oblique aerial photographs were provided for each tank commander, and plasticine models of every part of the eventual battle area were carefully prepared, the shelled zone being stencilled on them as the bombardment proceeded. To facilitate the movement of tanks over the battlefield a new system was made use of by which a list of compass bearings from well-defined points to a number of features in the enemy's territory was prepared, thus enabling direction to be picked up easily. The distribution of information was more rapid than it had been on previous occasions. Constant discussions between the Brigade and Battalion Reconnaissance officers led to a complete liaison; in fact, everything possible was done to make the tank operations a success; further, there was ample time to do it in.

The Fifth Army attack was to be carried out on well recognised lines; namely, a lengthy artillery preparation followed by an infantry attack on a large scale and infantry exploitation until resistance became severe, when the advance would be halted and a further organised attack prepared on the same scale. This methodical progression was to be continued until the exhaustion of the German reserves (at this time the German reserves totalled about 750,000 men), and moral created a situation which would enable a complete break-through to be effected.

The number of tanks allotted to the three attacking corps was as follows: seventy-two to each of the IInd and XIXth Corps, thirty-six to the XVIIIth, and thirty-six to be held in Army reserve. These were subdivided according to objectives, namely:

	IInd Corps	XIXth Corps	XVIIIth Corps
Black Line	16 tanks	24 tanks	12 tanks
Green Line	24 "	24 "	12 "
East of Green Line	8 "		
Corps Reserve	24 "	24 "	12 "

The Corps objectives and the allotment of tanks to divisions were as follows:

In the IInd Corps the 24th and 30th Divisions supported by the

18th Division were to attack on the right, and the 8th Division, supported by the 25th Division, on the left. The general objective of the operations was the capture of the Broodseinde ridge, and the protection of the right flank of the Fifth Army. The allotment of tanks to divisions was: twelve to the 30th Division, eight to the 18th Division, twenty-four to the 8th Division, and four to the 24th Division.

In the XIXth Corps the 15th Division was on the right and the 55th Division on the left, with the 16th and 36th Divisions in reserve. The objective of this Corps was to capture and hold a section of the enemy's third-line system known as the Gheluvelt-Langemarck line. Twenty-four tanks were allotted to each of the attacking divisions.

In the XVIIIth Corps the 39th Division was on the right and the 51st Division on the left, with the 11th and 48th Division in reserve. The main objective was the Green Line; but should this be successfully occupied the 51st Division was to seize the crossings of the River Steenbeek at Mon du Rasta and the Military Road, and establish a line beyond that river from which a further advance could be made on to the Gheluvelt-Langemarck line; the 39th Division on the left conforming by throwing out posts beyond the Green Line. Eight tanks were allotted to the 51st Division and sixteen to the 39th Division.

The dead level of Northern Flanders is broken by one solitary chain of hills, a crescent in shape, with its cusps as Cassel and Dixmude. From Cassel to Kemmel hill had been ours since 1914; to this the Messines-Wytschaete ridge was added, as we have seen in June 1917; now all that remained was the extension of this ridge northwards from about Hooge to Dixmude. The territory lying within the crescent was practically all reclaimed swamp land, including Ypres and reaching back as far as to St. Omer, both of which, a few hundred years ago, were seaports. All agriculture in this area depended on careful drainage, the water being carried away by innumerable dikes. So important was the maintenance of this drainage system considered that in normal times a Belgian farmer who allowed his dikes to fall into disrepair was heavily fined.

The frontage of attack of the Fifth Army extended from the Ypres-Comines canal to Wiltje cabaret. On the left the French were co-operating, attacking towards Houthulst forest, and on the right the Second Army was restricted to an all but passive artillery role. This frontage was flanked by two strong positions, the Polygonveld and Houthulst forest, which formed two bastions with a semi-circular ridge of ground as a curtain between them; in front of this low curtain ran a broad moat—

the valley of the Steenbeek and its small tributaries.

From the tank point of view the Third Battle of Ypres is a complete study of how to move thirty tons of metal through a morass of mud and water. The area east of the canal had, through neglect and daily shell fire, been getting steadily worse since 1914, but as late as June 1917 it was still sufficiently well drained to be negotiable throughout, by the end of July it had practically reverted to its primal condition of a vast swamp; this was due to the intensity of our artillery fire.

It must be remembered at this time the only means accepted whereby to initiate a battle was a prolonged artillery bombardment; sufficient reliance not as yet being placed in tanks on account of their liability to break down. (Breakdowns in the past had for the most part been due to bad ground, not defective mechanism.) The present battle was preceded by the longest bombardment ever carried out by the British Army, eight days counter-battery work being followed by sixteen days intense bombardment. The effect of this cannonade was to destroy the drainage system and to produce water in the shell-holes formed even before the rain fell. Slight showers fell on the 29th and 30th, and a heavy storm of rain on July 31.

A study of the ground on the fronts of the three attacking corps is interesting. On the IInd Corps front the ground was broken by swamps and woods, only three approaches were possible for tanks, and these formed dangerous defiles. On the XIXth Corps front the valley of the Steenbeek was in a terrible condition, innumerable shell-holes and puddles of water existed, the drainage of the Steenbeek having been seriously affected by the shelling. On that of the XVIIIth Corps front the ground between our front line and the Steenbeek was cut up and sodden. The Steenbeek itself was a difficult obstacle, and could scarcely have been negotiated without the new unditching gear which had been produced since the Battle of Messines. The only good crossing was at St. Julien, and this formed a dangerous defile.

Zero hour was at 3.50 a.m., and it was still dark when the tanks, which had by July 31 assembled east of the canal, moved forward behind the attacking infantry.

Briefly, the attack on July 31, in spite of the fact that there are fifty-one recorded occasions upon which individual tanks assisted the infantry, may be classed as a failure. On the IInd Corps front, because of the bad going, the tanks arrived late, and owing to the infantry being hung up, they were caught in the defiles by hostile artillery fire and suffered considerable casualties in the neighbourhood of Hooge.

THIRD BATTLE OF YPRES
July to November 1917.

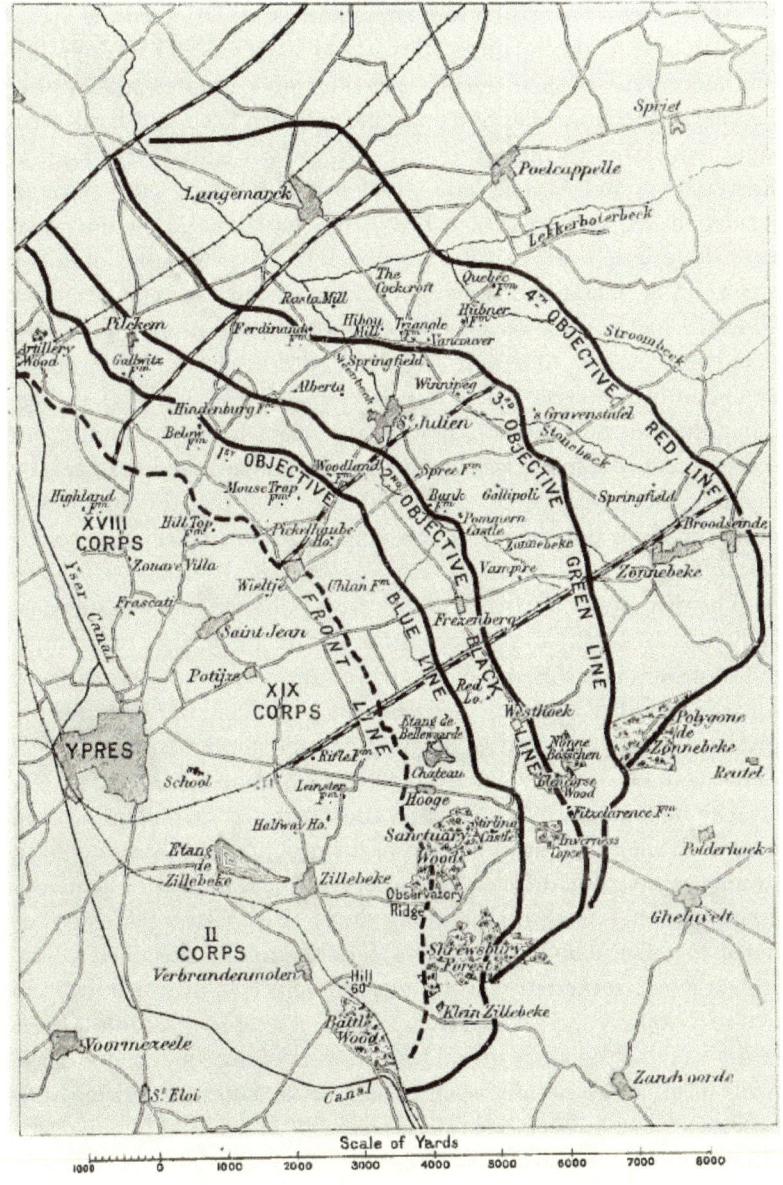

They undoubtedly drew heavy shell fire away from the infantry, but the enemy appeared to be ready to deal with them as soon as they reached certain localities and knocked them out one by one. On the XIXth Corps front they were more successful.

At the assault on the Frezenberg redoubt they rendered the greatest assistance to the infantry, who would have suffered severely had not tanks come to their rescue. Several enemy's counter-attacks were broken by the tanks, and Spree farm, Capricorn keep, and Bank farm were reduced with their assistance. On the XVIIIth Corps front at English trees and Macdonald's wood several machine guns were silenced; the arrival of a tank at Ferdinand's farm caused the enemy to evacuate the right bank of the Steenbeek in this neighbourhood. The attack on St. Julien and Alberta would have cost the infantry heavy casualties had not two tanks come up at the critical moment and rendered assistance. At Alberta strong wire still existed, and this farm was defended by concrete machine-gun emplacements with good dugouts. The two tanks which arrived here went forward through our own protective barrage, rolled flat the wire and attacked the ruins by opening fire at very close range, with the result that the enemy was driven into his dug-outs and was a little later on taken prisoner by our infantry.

The main lessons learnt from this day's fighting were—the unsuitability of the Mark IV tank to swamp warfare; the danger of attempting to move tanks through defiles which are swept by hostile artillery fire; the necessity for immediate infantry co-operation whenever the presence of a tank forced an opening, and the continued moral effect of the tank on both the enemy and our own troops.

The next attack in which tanks took part was on August 19, and in spite of the appalling condition of the ground, for it had now been steadily raining for three weeks, a very memorable feat of arms was accomplished. The 48th Division of the XVIIIth Corps had been ordered to execute an attack against certain strongly defended works, and, as it was reckoned that this attack might cost in casualties from 600 to 1,000 men, it was decided to make it a tank operation in spite of the fact that the tanks would have to work along the remains of the roads in place of over the open country. Four tanks were detailed to operate against Hillock farm, Triangle farm, Mon du Hibou, and the Cockcroft; four against Winnipeg cemetery, Springfield, and Vancouver, and four to be kept in reserve at California trench. The operation was to be covered by a smoke barrage, and the infantry were to follow

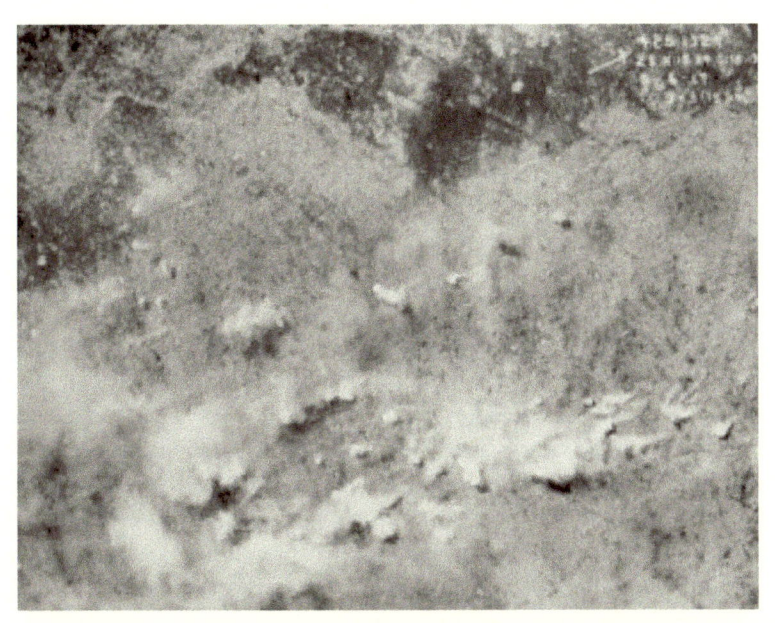

Ground Operated Over by Tanks During the Battle of Messines, Showing Preliminary Bombardment on June 5, 1917.

Ground Operated Over by Tanks in August 1917, During the Third Battle of Ypres.

the tanks and make good the strong points captured.

Eleven tanks entered St. Julien at 4.45 a.m., three ditched, and eight emerged on the St. Julien-Poelcappelle road, when down came the smoke barrage, throwing a complete cloud on the far side of the objectives; at 6 a.m. Hillock farm was occupied, at 6.15 a.m. Mon du Hibou was reduced, and five minutes later the garrison of Triangle farm, putting up a fight, were bayoneted. Thus, one point after another was captured, the tanks driving the garrisons underground or away, and the infantry following and making good what the tanks had made possible. In this action the most remarkable results were obtained at very little cost, for instead of 600 casualties the infantry following the tanks only sustained fifteen!

From this date on to October 9 tanks took part in eleven further actions, the majority being fought on the XVIIIth Corps front by the 1st Tank Brigade. On August 22 a particularly plucky fight was put up by a single tank. This machine became ditched in the vicinity of a strong point called Gallipoli, and, for sixty-eight hours on end, fought the enemy, breaking up several counter-attacks; eventually the crew, running short of ammunition, withdrew to our own lines on the night of August 24–25.

Of the attacks which were made with tanks in the latter half of September and the beginning of October, the majority took place along the Poelcappelle road, the most successful being fought on October 4. Of this attack the XVIIIth Corps Commander reported that "the tanks in Poelcappelle were a decisive factor in our success on the left flank"; and their moral effect on the enemy was illustrated by the statement of a captured German officer who gave as the reason of his surrender—"There were tanks—so my company surrendered—I also."

It is almost impossible to give any idea of the difficulty of these latter operations or of the "grit" required to carry them out. Roads, if they could be called by such a name at all, were few and far between in the salient caused by the repeated attacks during the battle. This salient had a base of some 20,000 yards and was only 8,000 deep at the beginning of October, at which date the enemy could still obtain extensive observation over it from the Passchendaele ridge. The ground in between these roads being impassable swamps, all movement had to proceed along them, consequently they formed standing targets for the German gunners to direct their fire on. One night, at about this period of the battle, a tank engineer officer was instructed to proceed to Poelcappelle to superintend the demolition of some tanks which were blocking the road near the western entrance to the village. His description of it at night-time is worth recording.

> I left St. Julien in the dark, having been informed that our guns were not going to fire. I waded up the road, which was swimming in a foot or two of slush, frequently I would stumble into a shell-hole hidden by the mud. The road was a complete shambles and strewn with debris, broken vehicles, dead and dying horses and men. I must have passed hundreds of them as well as bits of men and animals littered everywhere. As I neared Poelcappelle our guns started to fire: at once the Germans replied, pouring shells on and around the road, the flashes of the bursting shells were all round me. I cannot describe what it felt like, the nearest approach of a picture I can give is that it was like standing in the centre of the flame of a gigantic Primus stove. As I neared the derelict tanks, the scene became truly appalling: wounded men lay drowned in the mud, others were stumbling and falling through exhaustion, others crawled and rested themselves up against the dead to raise themselves a little above the slush. On reaching the tanks I found them surrounded by the dead and dying; men had crawled to them for what shelter they would afford. The nearest tank was a female, her left sponson doors were open, out of these protruded four pairs of legs, exhausted and wounded men had sought refuge in this machine, and dead and dying lay in a jumbled heap inside.

Whatever history may record of the Third Battle of Ypres, one fact certainly will not be overlooked or forgotten, namely: that men who could continue for three months to attack under the conditions which characterised this most terrible Battle of the war must indeed belong to an invincible stock.

CHAPTER 16
Tank Mechanical Engineering

The organisation of the "mechanical engineering" side of the Tank Corps constituted the backbone of the whole formation, for on its efficiency depended the efficiency of the fighting units in as high a degree as the fighting efficiency of a cavalry regiment depends on its horse-mastership.

In this chapter it is not intended to follow the growth of this organisation in detail, but rather to look back on its evolution as a whole, and then to enter upon a few particulars of the work accomplished by it. Before doing so it must be clearly understood that the mechanical engineering side of the Tank Corps was as much a product of this corps as the fighting organisation itself, as there was in the army no definite Mechanical Engineering Department to draw inspirations from. The nearest was the R.A.S.C., but a very wide gap separated the R.A.S.C. system from that adopted in the Tank Corps; both indeed dealt with petrol engines, but the tank and its requirements are as distinct from the lorry as the lorry is from the aeroplane—another mechanical weapon.

Generally speaking, the experience of the engineering side of the Tank Corps, during the two years following its inception in August 1916, has been that the most efficient organisation depends upon the maintenance of two simple principles, namely:

(1) No repairs to be carried out in the field—*i.e.* by fighting units.

(2) All maintenance to be carried out by the crews of the machines themselves.

When the Tank Corps was first formed, each Company of Tanks was provided with its own workshops. At the end of 1916 Company Workshops were abolished and Battalion Workshops were formed. Towards the end of 1917, after much consideration had been given to the question, Battalion Workshops were abolished and merged into Brigade Workshops, while a small number of skilled workshop men were left with each company. In 1918 it was realised that the gradual withdrawal of special workshop facilities from the Company organisation to the brigade had resulted in a considerable improvement in the skill and ability of the tank crews themselves in the maintenance of their tanks. It was decided, therefore, to go one step further, and not only withdraw all Brigade Workshops into a central organisation known as the Central Workshops, but also to withdraw the special workshop men from the companies, while tank crews themselves were made entirely responsible for the maintenance of their machines.

In this way it was possible to draw a clear line between maintenance (*i.e.* the replacing of damaged parts, which was done entirely by the crews) and repairs (*i.e.* the mending of broken parts, which was done entirely by the Central Workshops). At this time the argument was frequently heard that a man who uses a machine should be able to repair it, and that, if all repair work is done by a different organisation from the one which actually fights the machine, there will be a serious loss of mechanical efficiency. This idea was based upon a misconception of the difference between the functions of repairs and maintenance. On the contrary, it was found that the efficiency of the crews increased several hundred *per cent.* after the crews themselves were made responsible for the maintenance of their machines. To carry out this system it is, however, necessary that stores and spare parts should be readily available in the field; this entails an intelligent system of Advanced Stores.

One very great advantage of this centralisation of repair work is the considerable saving in man-power effected by employing all skilled men exclusively on one particular job. As an example, broken unions of petrol pipes commonly occur in all petrol engines, and if a small unit workshop exist, the brazing out and repair of such broken unions can be carried out there. In order to do this a coppersmith must be kept at the unit workshop, and only part of his time will be employed in this work of brazing petrol unions. If now, however, the unit workshops are abolished, and all broken unions, from every unit, are sent back to a Central Workshop for repair, there is a sufficient amount of work of this description to keep one man, or possibly two or three, fully employed all their time. These men become absolute experts in brazing broken unions, and before very long can do in a few minutes a job of this sort which would take a coppersmith with the unit workshop considerably longer.

The complete organisation for the maintenance and repair of tanks can be briefly described by tracing the itinerary of a tank from the day it left the manufacturer until the day it was received at the Central Workshops for repair.

From the manufacturer the tank was first sent to the tank testing ground at Newbury, which was manned and administered by No. 20 Squadron, R.N.A.S. From here it was sent to Richborough and shipped across the channel by channel ferry and received by another detachment of No. 20 Squadron at Havre. From Havre it was sent to the Bermicourt area, and after being put through further tests was handed over to the Central Stores. The Central Stores were situated at the village of Erin on the Hesdin-St.-Pol railway, and consisted of some seven acres of railway sidings and some six acres of buildings. These stores were built in 1917, and at first included the Central Workshops; in 1918, however, these workshops were installed at Teneur, about a mile and a half away, and covered some twenty acres of ground.

From the Central Stores tanks were issued to battalions, and after repair at the Central Workshops were received again at these stores for reissue as they were required.

As the battalions carried out all their own mechanical maintenance, Advanced Stores were instituted, these being sent out from the Central Stores into the forward area immediately behind the front to be attacked by the tanks. These stores were organised on a very mobile footing and proved invaluable in all battles since October 1917.

Besides the moving forward of these Advanced Stores, Tank Field Companies, originally known as Salvage Companies, were despatched from the Central Workshops to the battle areas. The duty of these companies was to take over from the fighting units all damaged tanks, such as those knocked out by the enemy's artillery fire; they were, in fact, the clearers of the battlefield so far as tanks were concerned. Apart from salving complete tanks an immense quantity of other material was reclaimed, such as 6-pounder guns, machine guns, ammunition, tools, track plates, gears, transmissions, and engine parts, etc., which in the two years of the existence of these companies totalled in value several millions of pounds.

The work carried out by the Tank Field Companies was particularly dangerous, and many casualties amongst their personnel occurred. In the actual reclaiming of machines or parts they were constantly under shell fire, and the actual carrying to and fro of the material made use of required great physical strength, since the ground to be traversed was frequently a mass of shell-holes; incidentally a great deal of work had to be done at night since many of the machines to be salved were frequently situated in full view of the enemy.

From the Tank Field Companies the salved machines were sent to the Central Workshops at Teneur. Here they were repaired, and this work entailed considerably more skill and labour than the initial assembly of the machines in the home factories on account of the shattered and burnt-out condition many of these machines were reduced to. Much of this repair work was carried out by Chinese labour and at these shops over 1,000 Chinese were quartered for work, schools being instituted for them so that mechanics, fitters, etc., could be trained.

Besides testing and repairing machines much other work was carried out at the Central Workshops—"gagget" making, experiments, making good minor deficiencies of manufacture and generally improving the machines. This work frequently consisted in "panic" orders, such as the sledge and fascine making for the Battle of Cambrai.

One hundred and ten tank sledges and 400 tank fascines, bundles of wood which will be mentioned later, when dealing with the Battle of Cambrai, were ordered on October 24, 1917. The former required some 3,000 cubic feet of wood, weighing 70 tons, and the latter 21,500 ordinary fascines, representing some 400 tons of brushwood and over 2,000 fathoms of chain to hold them together. This order came on the top of a particularly strenuous period following the tank operations round Ypres. At the same time another order for the overhaul and repair of 127 machines was made.

Owing to the limited amount of time allowed for the transport of material from the base ports to the Central Workshops and from the Central Workshops to the forward area, extensive use was made of lorries. From November 10 to the 25th, twenty-eight lorries engaged on this work covered a total of 19,334 miles, averaging 690 miles per lorry, while three box cars averaged 1,242 miles each.

The part played by the 51st Chinese Labour Company, attached to the Workshops, materially contributed to the work being duly completed in time for the battle. Owing to the necessity for secrecy the personnel of the Workshops were without knowledge of the immediate urgency of the work they were engaged on. In spite of this all ranks worked with the utmost enthusiasm, accomplishing the task in the required time. During these three weeks the Central Workshops were working 22½ hours out of every 24 without a break, and had it not been for the "grit" displayed by all ranks the Battle of Cambrai could not have been fought, and without this battle the whole course of the war might have been changed; for it was the Battle of Cambrai, as we shall shortly see, which demonstrated the full power of the tank and which placed it henceforth in the van of every battle.

CHAPTER 17
The Third Battle of Gaza

As a result of the repulse sustained by the British forces at the Second Battle of Gaza in April 1917, the troops operating were withdrawn from their exposed position and the Tank Detachment was concentrated in a fig grove some 2,000 yards west of Sheikh Nebhan, at which place it was later on reinforced by three Mark IV machines.

A new plan of operations was drawn up in which the Turkish defences from Outpost hill to Ali El Muntar, which had resisted the combined onslaught of several divisions, was to be turned by an extensive flanking movement west of Gaza. This operation was to take place in conjunction with an attack on Beersheba.

The general plan of attack was as follows:

(1) The Australian Corps and Desert Column were to operate from Beersheba north-westwards to Hareira.

(2) Several mounted and dismounted divisions were to operate around Hareira and Gaza.

(3) A composite force of French, Italians and West Indian troops was to demonstrate by raids in the vicinity of Outpost hill.

(4) The XXIst Corps was to attack the enemy's defences between Umbrella hill and the sea.

It is with the last of these four operations that this chapter is concerned. For this attack the force detailed consisted of the 54th Division, the Indian Cavalry Division, and the Tank Detachment—eight machines.

The attack of the 54th Division was divided into four phases—Blue, Red, Green, and Yellow.

The Tank Detachment left Deir El Belah on the night of October 22–23, 1917, for the beach near Sheikh Ajlin. From here a thorough reconnaissance of the area of operations was carried out on horseback and by drifter, after which this area was divided into tank sectors. To these tanks were distributed as follows:

(1) *156th Infantry Brigade.*—Objective—Umbrella hill north-westwards to the eastern portion of El Arish redoubt. No. 1 tank was to support the infantry in their attack on El Arish redoubt, deposit R.E. stores, attack Magdhaba trench and cover the infantry consolidation.

(2) *163rd Infantry Brigade.*—Objective—El Arish redoubt northwards to south of Zowaiid trench. No. 2 tank was to attack El Arish redoubt, then Island wood, deposit R.E. stores, and on the arrival of the infantry take a southerly route and capture Crested rock.

(3) *161st and 162nd Infantry Brigades.*—Objectives—Zowaiid trench to Sea Post and the Cricket valley northwards to Sheikh Hasan and 500 yards beyond. For this operation four tanks were allotted.

No. 3 tank to attack Zowaiid trench, move north and capture Rafa redoubt, rally, then proceed along Rafa trench, deposit R.E. material, capture Yunus trench, attack Belah trench until the arrival of infantry, thence proceed to Sheikh Hasan, deposit more R.E. stores, and return to Sheikh Ajlin.

No. 4 tank to attack Rafa redoubt, co-operate with No. 3 tank against Belah trench, make for Sheikh Hasan, deposit R.E. material, attack A.6 (an isolated Turkish trench to the north-east), and hold this until the infantry had consolidated it.

No. 5 tank to capture Beach Post, co-operate in the assault on Cricket redoubt, attack Sheikh Hasan in advance of the infantry, and deposit R.E. material.

No. 6 tank to capture Sea Post, crush down the wire as far as Beach Post, assault Gun hill, proceed to Sheikh Hasan, then deposit R.E. stores and capture the Turkish post A.5.

(4) *Reserve Tanks.*—Nos. 7 and 8 tanks were to be held in reserve north-east of Sheikh Ajlin; from there they were to follow up the attack and replace any disabled machine.

In all, the above six first-line machines had twenty-nine objectives to attack! That this could have been accomplished successfully would have demanded a miracle; it was consequently foredoomed to failure.

The first phase of the attack was to consist in an infantry assault protected by a creeping barrage; during this phase the tanks were to move to their starting-points and be ready to advance at 3 o'clock on the morning of November 2.

In order to ensure complete co-operation between the Tank Detachment and the infantry, tank officers and other ranks were attached to infantry brigades for ten days prior to the battle. As in France, this system resulted in the greatest benefit to both infantry and tank personnel alike.

The first phase of the battle opened at 11 o'clock on the night of November 1–2, the 156th Infantry Brigade assaulting Umbrella hill. To this attack the enemy did not respond immediately, but when he did, he opened a heavy artillery fire all down his front which endangered the forward movement of the tanks to their starting-points, which, in spite of this, they reached half an hour before the second zero hour.

It had been hoped to make every use of the full moon which rose in the early evening, but the smoke, resulting from the battle, and a dense haze restricted vision so completely that the tanks had to move forward on compass bearings. At 3 a.m. a heavy barrage was opened on the enemy's front-line system, behind which the tanks, followed by the infantry, moved forward. No difficulty was experienced in dealing with the first objective, as the Turks were evidently taken quite by surprise. Under cover of the barrage our troops pushed on till they approached their second objective, when the enemy's fire began to make itself felt. Along the coastline the attack proceeded according to programme, all objectives, including Sheikh Hasan, being taken. The 161st and 163rd Brigades encountered considerable opposition at Rafa trench, Island wood, Crested rock, Gibraltar, and north of El Arish redoubt. Briefly the operations carried out by the eight tanks of the detachment were as follows:

No. 1 tank successfully attacked El Arish redoubt and was penetrating the maze of trenches beyond, when owing to the darkness it was ditched; its crew then joined the infantry.

No. 2 tank assaulted El Arish redoubt and met a similar fate; eventually it received a direct hit which broke its right track; its crew also joined the infantry.

No. 3 tank attacked Rafa redoubt, and later on, losing direction in the mist, rallied.

No. 4 tank assaulted Rafa Junior, Yunus and Belah trenches, and after depositing its R.E. material rallied.

No. 6 tank captured Sea Post, moved along the enemy's trench, crushing down the wire as far as Beach Post, attacked Cricket Redoubt, Gun Hill, and Tortoise Hill, and reaching Sheikh Hasan deposited its R.E. stores. Shortly after this it moved forward to attack A.5, but breaking a track it had to be abandoned.

Nos. 7 and 8 tanks received instructions at 4 a.m. to support the infantry attack on El Arish redoubt, and to proceed in this direction with R.E. material. These machines were loaded up with empty sandbags on the roof, these caught fire, probably through the heat of the exhaust pipe, both tanks went out of action.

On the whole the tank operations during the Third Battle of Gaza were of assistance to the infantry. All tanks, except one, reached their first objectives; four reached their second, third, and fourth, and one reached its fifth objective. Of the eight machines operating five were temporarily disabled. Casualties in personnel were very light—only one man being killed and two wounded.

The Third Battle of Gaza closed the tank operations with the Army of Palestine, and though the damaged machines were repaired and put into fighting trim they were not used again. In order to overcome the great difficulty in rounding up, by means of cavalry, the rear-guard detachments of the retiring Turkish Army, a mission was sent to France to obtain, if possible, a number of Whippet machines.

This mission reached Tank Corps Headquarters in France on March 21, the day the German offensive was launched. All hope of procuring these machines consequently vanished. The Tank Detachment, therefore, handed over their machines to the Ordnance Department at Alexandria and returned to England.

The tank operations in Sinai and Palestine conclusively proved that tanks could be employed almost anywhere in desert regions, and all that they required were certain improvements in mechanism and changes in design.

What success the Tank Detachment won during its two years in the East was due to the determination and fine fighting spirit displayed by its officers and men, who laboured under the greatest difficulties, not the least being the entire lack of knowledge displayed by the other arms in the limitations of tanks and their tactical employment. When all is said and done and every criticism rounded off, success with tanks in battle is as much a matter of co-operation, that is, unity of action of all arms combined, as it is of mechanical fitness. This can only be attained by constant practice in combined exercises on the training area long prior to the battle even being thought of.

CHAPTER 18
Origins of the Battle of Cambrai

The battles of 1914 were primarily infantry battles based on the power of the rifle, and it was not until 1915 that the quick-firing gun and the heavy howitzer began to replace the rifle as the reducers of resistance to the infantry attack.

In 1915, as far as the British Army is concerned, it may be said that artillery was generally looked upon as an adjunct to the infantry. This idea died hard, and it was not until the Battle of the Somme was half way through that it became apparent that it was no longer a question of guns co-operating in the infantry advance, but of the infantry itself co-operating with the artillery bombardments. In other words, the limit of the infantry advance was the limit of the range of the guns, particularly the 18-pounders. This meant that no penetration of a greater depth than about 4,000 yards could be effected unless a second echelon of guns and infantry was launched and brought into action simultaneously with or just before the first assault had reached the range limit of the guns which were supporting it.

Two factors prohibited this from being done; the first was the heavily shelled area which usually extends over the whole 4,000 yards of the initial advance; the second, the great difficulty of keeping the second echelon of guns supplied with sufficient ammunition to maintain the second barrage. The shelled area not only prohibited the movement of guns for days, but produced such exhaustion on the second echelon of infantry crossing it, that by the time it caught up with the first echelon its men were too fatigued to continue the attack.

By some it was hoped that the tank would partially overcome this difficulty if it were brought into action on the area over which the 18-pounder barrage was beginning to fail, and by producing a local barrage of its own, by means of its machine guns, that it would cover any advance from 4,000 yards onwards.

This idea, though perfectly sound in itself, was doomed to failure as long as the conditions of ground produced by artillery fire rendered it impossible to support these tanks by infantry in fighting condition.

The first solution to this problem was to cease using heavy artillery on ground to be traversed by the infantry attackers. This, however, is at best but a half measure, for though, in the present phase of the war, co-operation between infantry and tanks was of vital importance, this co-operation could not be maintained for long if one arm has to rely on its muscular power, whilst the other relies on petrol as its motive force. At best, the advance of 4,000 to 6,000 yards will be extended to 8,000 or 12,000 yards, when the endurance of the infantry will reach its limit and the advance automatically cease. This is not sufficient, for in a war such as was being waged in 1917 (a trench war), in order to beat an enemy, the first necessity was to prevent him using his spade. This can only be done by maintaining a continuous, if comparatively slow advance, that is, by replacing muscle by petrol as the motive force. This means the creation of a mechanical army.

In August 1917 the Tank Corps fully realised that the creation of such an army, even on a very small scale, would take at least a year; further, that its creation depended on the value of the tank being fully recognised by those who could create such an army, and upon this army being used in a suitable area of operations. Consequently, the first thing to do was to discover a suitable tank area; the second, to hold a tactical demonstration on it with tanks so as to convince the General Staff of their power and value. These steps it was felt would have to be taken before the petrol engine would be accepted as the motive force of the modern battle.

The selection of a theatre of operations depends on the objective to be gained; the gaining of the objective on the breaking down of the enemy's resistance. Consequently, the weapon which will most speedily overcome this resistance must be considered first, and the area of attack in the theatre of operations chosen must be selected as far as possible with reference to its powers.

In the present instance we find that the chief resistance to our infantry advance comes from the enemy's machine guns. We dare not concentrate all our artillery on these, for if we do, we shall release his guns, which, free, can put up a stronger resistance than his machine guns on account of their superior range. Further, whilst by sound and flash ranging and aeroplane observation, we can discover his main gun positions, no means have yet been discovered whereby his machine guns can be located other than by advancing on them and risking casualties. Tanks must, therefore, be employed to do this in order to clear the way for the infantry advance. Consequently, if sufficient tanks are forthcoming, in order to guarantee a decisive success, it is no longer a question of the tank as a spare wheel to the car, in case of an unforeseen puncture in our operations, but as the motive force of the car itself, the infantry being merely its armed occupants; without which the car is valueless.

The area of operations selected must firstly be suitable to the rapid movement of tanks, and secondly, unsuitable to hostile anti-tank defences. Further, it should be chosen with reference to the tactical characteristics of this arm. Once chosen, all other weapons should be deployed and employed to facilitate the advance of the tank, because it is to be used as the chief maintainer of infantry endurance, and it is the infantry man with his machine gun and bayonet who is going, for some time to come, to decide the battle.

Such were the views held in the Tank Corps at the opening of the Third Battle of Ypres, and the following extract taken from a paper written on June 11, 1917, is not only of interest but prophetic of future events:

> If we look at a layered map of France we can at once put our finger on the area to select. It lies between the Scarpe and the Oise, the Flanders swamps in the north and the Ardennes in the south-east. It was down this funnel of undulating country that the Germans advanced in 1914, and it is up it that they will most likely be driven if strategy is governed by ground and tactics by weapons.
>
> The main area suitable for tank operations having been fixed upon

by the Tank Corps, the next requirement was to select a definite objective, the attack against which would draw the enemy's reserves towards it and so relieve the pressure which was being exerted against the Fifth Army at Ypres. Two localities were considered, St. Quentin and Cambrai. The first was opposite the French area, the second opposite the British.

The suggestion put forward as regards the St. Quentin operation was abandoned on account of difficulties arising out of a British force operating in the French area; it must be remembered that at this time no real unity of command existed in France.

The Cambrai project consisted in a surprise raid, the duration of which would be about twenty-four hours. The whole operation may be summed up in three words, "Advance, Hit, Retire." Its object was to destroy the enemy's personnel and guns, to demoralise and disorganise his fighting troops and reserves, and not to capture ground or to hold trenches. It was further considered that such an operation would interrupt his *roulement* of reserves and make the enemy think twice as to replacing fresh divisions by exhausted and demoralised units in those parts of his line which were not included in his battle front. Further, it would confuse him as to the decisive point of attack; for any day one of these raids might be followed by a strong offensive.

The actual area of operations selected was the re-entrant formed by the L'Escaut or St. Quentin Canal between the villages of Ribecourt, Crèvecœur, and Banteux. The going in this area was excellent; further, the area to be raided contained several fair-sized villages and important ground, and was well limited by the canal, which not only made a rapid reinforcing of the area in the bend difficult, but completely limited the tank objectives.

The plan of attack was a threefold one:

(1) To scour the country between Marcoing, Masnières, Crèvecœur, Le Bosquet, Banteux.

(2) To form an offensive flank between Le Bosquet and Ribecourt.

(3) To form an offensive flank against Banteux.

The attack was to be launched at dawn, the first line of tanks making straight for the enemy's guns, which before, and as the tanks approached them, were to be bombed by our aeroplanes. The second and third lines of tanks were to follow, whilst our heavy guns commenced counter-battery work and the shelling of the villages and bridges along the canal. The essence of the entire operation was to be surprise coupled with rapidity of movement. The spirit of such an enterprise is audacity, which was to take the place of undisguised preparation.

It must be realised that both the St. Quentin and Cambrai projects were the home product of the Tank Corps, and they did not emanate from higher authority, which, when approached, was unable to sanction either. In spite of this, steps were taken to reconnoitre the Cambrai area, and for this purpose both the Brigadier-General commanding the Tank Corps and the 3rd Tank Brigade Commander visited Sir Julian Byng, the Third Army Commander, at his Headquarters in Albert. Though it is not known whether the Third Army Commander had already considered the possibilities of an offensive on the front of his Army, in September it would appear that he approached G.H.Q. on the subject, with the result that still no action outside the Ypres area could be considered, anyhow for the present.

CHAPTER 19
The Battle of Cambrai

On October 20, the project, which had been constantly in the mind of the General Staff of the Tank Corps for nearly three months and in anticipation of which preparations had already been undertaken, was approved of, and its date fixed for November 20.

The battle was to be based on tanks and led by them. There was to be no preliminary artillery bombardment; the day the Tank Corps had prayed for, for nearly a year, was at last fixed, and its success depended on the following three factors:

(1) That the attack was a surprise.

(2) That the tanks were able to cross the great trenches of the Hindenburg system.

(3) That the infantry possessed sufficient confidence in the tanks to follow them.

The following difficulties had to be overcome before these requirements could be met. The tanks, on October 20, were scattered over a considerable area: some were at Ypres, others near Lens, and others at Bermicourt. These would all have to be assembled not at suitable entraining stations, as is usually the case, but at various training areas so that co-operative training with the infantry could take place. This was of first importance, for success depended as much on the confidence of the infantry in the tanks as on the surprise of the attack. At these training centres, tanks would have to be completely overhauled and fitted with a special device to assist them in crossing the Hindenburg trenches, which were known, in many places, to be over 12 ft. wide, and the span of the Mark IV machine was only 10 ft. This device consisted in binding together by means of chains some seventy-five ordinary fascines, thus making one tank fascine, a great bundle of brushwood 4½ ft. in diameter and 10 ft. long; this bundle was carried on the nose of the tank and, when a large trench was encountered, was cast into it by pulling a quick release inside the tank. As already described these tank fascines and the "fitments" necessary to fix and release them were made by the Tank Corps Central Workshops.

Before the infantry assembled for training a new tactics had to be devised, not only to meet the conditions which would be encountered but to fit the limitations imposed upon the tank by it being able to carry only one tank fascine. Once this fascine was cast it could not be picked up again without considerable difficulty.

Briefly, the tactics decided on were worked out to meet the following requirements: "To effect a penetration of four systems of trenches in a few hours without any type of artillery preparation." They were as follows:

Each objective was divided up into tank section attack areas, according to the number of tactical points in the objective, and a separate echelon, or line, of tanks was allotted to each objective. Each section was to consist of three machines—one Advanced Guard tank and two Infantry tanks (also called Main Body tanks); this was agreed to on account of there not being sufficient tanks in France to bring sections up to four machines apiece.

The duty of the Advanced Guard tank was to keep down the enemy's fire and to protect the Infantry tanks as they led the infantry through the enemy's wire and over his trenches. The allotment of the infantry to tanks depended on the strength of the objective to be attacked, and the nature of the approaches; their formation was that of sections in single file with a leader to each file. They were organised in three forces: trench clearers to operate with the tanks; trench stops to block the trenches at various points, and trench supports to garrison the captured trench and form an advanced guard to the next echelon of tanks and infantry passing through.

The whole operation was divided into three phases: the Assembly, the Approach, and the Attack. The first was carried out at night time and was a parade drill, the infantry falling in behind the tanks on tape lines, connected with their starting-points by taped routes. The

Approach was slow and orderly, the infantry holding themselves in readiness to act on their own initiative. The Attack was regulated so as to economise tank fascines; it was carried out as follows. The Advanced Guard tank went straight forward through the enemy's wire and, turning to the left, without crossing the trench in front of it, opened right sponson broadsides. The Infantry tanks then made for the same spot: the left-hand one, crossing the wire, approached the trench and cast its fascine, then crossed over the fascine and, turning to the left, worked down the fire trench and round its allotted objective; the second Infantry tank crossed over the fascine of the first and made for the enemy's support trench, cast its fascine, and, crossing, did likewise. Meanwhile the Advanced Guard tank had swung round, and crossing over the fascines of the two Infantry tanks moved forward with its own fascine still in position. When the two Infantry tanks met, they formed up behind the Advanced Guard tank and awaited orders.

In training the infantry the following exercises were carried out:

(1) Assembly of infantry behind tanks.

(2) Advance to attack behind tanks.

(3) Passing through wire crushed down by tanks.

(4) Clearing up a trench sector under protection of tanks.

To enable them to work quickly in section single files and to form from these into section lines, a simple platoon drill was issued, and it is interesting to note that this drill was based on a very similar one described by Xenophon in his "*Cyropædia*" and attributed to King Cyrus (*circa* 500 B.C.).

Whilst training was being arranged by the Tank Corps General Staff the Administrative Staff was preparing for the railway concentration, which was by no means an easy problem.

The difficulties of concentrating a large number of tanks in the area of operations was accentuated by the dispersion of the Tank Corps and the shortage of trucks; this shortage was made good by collecting a number of old French heavy trucks; these, however, did not prove at all satisfactory as they were too light. In spite of these difficulties the whole of the units of the Tank Corps were concentrated in their training areas by November 5.

In order to make the most of the available truckage and the time attainable for infantry training, it was decided to concentrate three-quarters of the whole number of tanks to be used, *i.e.* twenty-seven train loads, at the Plateau station by November 14 (Z-6 days); to move these to their final detraining stations on Z-4, Z-3, and Z-2 days; and to move the remaining quarter, *i.e.* nine train loads, from the training areas to the detraining stations on Z-5 day. The detraining stations selected were Ruyaulcourt and Bertincourt for the 1st Brigade, Sorel and Ytres for the 2nd Brigade, and Old and New Heudicourt for the 3rd Brigade. At all these stations detraining ramps and sidings were built or improved. In all, thirty-six tank trains were run, and except for two or three minor accidents the move was carried out to programme. This was chiefly due to the excellent work of the Third Army Transportation Staff.

Supply arrangements were divided under two main headings: supply by light railways and supply in the field by supply tanks. The main dumps selected were at Havrincourt wood for the 1st Brigade, Dessart wood for the 2nd, and Villers Guislain and Gouzeaucourt for the 3rd Brigade. A few of the items dumped were 165,000 gallons of petrol, 55,000 lb. of grease, 5,000,000 rounds of S.A.A., and 54,000 rounds of 6-pounder ammunition. Without the assistance of the light railways this dumping would hardly have been possible. On November 30 a S.O.S. call for petrol was made on Ruyaulcourt. A train was loaded, despatched 3½ miles, and the petrol delivered in just under one hour. This is a fair example of the magnificent work consistently carried out by the Third Army light railways during the Battle of Cambrai.

The Third Army plan of operations was as follows:

(1) To break the German defensive system between the canal St. Quentin and the canal Du Nord.

(2) To seize Cambrai, Bourlon wood, and the passages over the River Sensée.

(3) To cut off the Germans in the area south of the Sensée and west of the canal Du Nord.

(4) To exploit the success towards Valenciennes.

This operation, for its initial success, depended on the penetration of all lines of defences, including the Masnières-Beaurevoir line, which in its turn depended on the seizing of the bridges at Masnières and Marcoing.

The force allotted for this attack was—two corps of three infantry divisions each; the Tank Corps of nine battalions—378 fighting tanks and 98 administrative machines; a cavalry corps, and 1,000 guns.

The attack was to be carried out in three phases. In the first the infantry were to occupy the line Crèvecœur, Masnières, Marcoing, Flesquières, canal Du Nord; the leading cavalry division was then to push through at Masnières and Marcoing, capture Cambrai, Paillencourt, and Pailluel (crossing over the River Sensée), and move with its right on Valenciennes; whilst this was in progress the IIIrd Corps, which formed the right wing of the Third Army, was to form a defensive flank on the line Crèvecœur, La Belle Etoile, Iwuy; the cavalry were then to cut the Valenciennes-Douai line and so facilitate the advance of the IIIrd Corps in a north-easterly direction. The second and third phases were to be carried out by the IVth Corps, which formed the left wing of the Third Army, and were to consist firstly in opening the Bapaume-Cambrai road and occupying Bourlon and Inchy, and secondly, in opening the Arras-Cambrai road and advancing on the Sensée canal and so to cut off the German forces west of the canal Du Nord.

The ground to be fought over consisted chiefly in open, rolling downland, very lightly shelled, and consequently most suitable to tank movement. The main tactical features were the two canals which practically prohibited the formation of tank offensive flanks and so strategically were a distinct disadvantage to what was meant to be a decisive battle. Between these two canals were two important features—the Flesquières-Havrincourt ridge and Bourlon hill. A third very important feature, known as the Rumilly-Seranvillers ridge, ran parallel to and north of the St. Quentin canal between Crèvecœur and Marcoing; without the occupation of this ridge a direct attack from the south on Bourlon hill could only take place under the greatest disadvantage.

The German defences consisted of three main lines of resistance and an outpost line: these lines were the Hindenburg Line, the Hindenburg Support Line, and the Beaurevoir-Masnières-Bourlon line, the last being very incomplete. The trenches for the most part were sited on the reverse slopes of the main ridges, and consequently direct artillery observation on them from the British area was impossible. They were protected by immensely thick bands and fields of wire arranged in salients so as to render their destruction most difficult. To have cut these bands by artillery fire would have required several weeks bombardment and scores of thousands of tons of ammunition.

The weather had been throughout November fine and foggy, so much so that aeroplane observation had been next to impossible. This foggy weather greatly assisted preparatory arrangements by securing them from observation.

The artillery preparations were as follows:—Over 1,000 guns of various calibres were concentrated in the Third Army area for the attack. None of these, however, were permitted to register before zero hour. Briefly the following was the artillery programme from zero hour on.

At zero the barrage was to open on the enemy's outpost line; it was to consist of shrapnel and H.E. mixed with smoke shells. It was to move forward by jumps of approximately 250 yards at a time, standing on certain objectives for stated periods. Simultaneously with this jumping barrage smoke screens were to be thrown up on selected localities, notably on the right flank of the IIIrd Corps and on the Flesquières ridge; counter-battery work was to open and special bombardments on prearranged localities such as bridgeheads, centres of communication, and roads likely to be used by the German reserves were to take place.

Tank Corps reconnaissances were started as early as secrecy would permit, but it was not until a few days before November 20 that commanders were allowed to reconnoitre the ground from our front-trench system. Meanwhile at the Plateau station tanks were tuned up and tank fascines fixed. All detrainments were carried out by night, the tanks being moved up to their position of assembly under cover of darkness. These positions were: Villers Guislain and Gouzeaucourt for the 3rd Brigade, Dessart wood for the 2nd Brigade, and Havrincourt wood for the 1st Brigade. At these places, tanks were carefully camouflaged.

The allotment of tanks to infantry units is given in the table. opposite Besides these, each Brigade had eighteen supply tanks or gun carriers and three wireless-signal tanks. Thirty-two machines were specially fitted with towing gear and grapnels to clear the wire along the cavalry lines of advance; two for carrying bridging material for the cavalry and one to carry forward telephone cable for the Third Army Signal Service. The total number of tanks employed was 476 machines.

On the night of November 17–18 the enemy raided our trenches in the vicinity of Havrincourt wood and captured some of our men, and, from the documents captured during the battle, it appears that these men informed the enemy that an operation was impending; time wherein the Germans could make use of this was, however, so limited that the warning of a possible attack only reached the German firing line a few minutes before it took place.

The following night, that of the 19th–20th, was broken by a sharp burst of artillery and trench-mortar fire which died away in the early morning, and at 6 a.m. all was still save for the occasional rattle of a machine gun. A thick mist covered the ground when at 6.10 a.m.,

DISTRIBUTION OF FIGHTING TANKS

Tank Bde.	Tank Bn.	Corps.	Divn.	Bde.	No. of Tanks.	Objective.	Exploit towards—	Remarks.
3rd Bde.	C	IV	12th	35th	24 / 4*	Blue	Crèvecœur	Number of tanks used for exploitation varied according to condition of units after gaining Blue and Brown Lines.
,,	,,	,,	,,	37th	12 / 2*	Brown		
,,	F	,,	,,	36th	24 / 4*	Blue	Masnières	
,,	,,	,,	,,	36th	12 / 2*	Brown		
,,	I	,,	20th	61st	18 / 3*	Vacquerie	Crèvecœur	
,,	,,	,,	,,	,,	12 / 2*	Blue		
,,	,,	,,	,,	62nd	6 / 1*	Brown		
,,	A	,,	,,	60th	18 / 3* / 3*	Blue	Canal, Masnières to Marcoing.	
,,	,,	,,	,,	,,	6 / 1*	Brown		
,,	,,	,,	29th	,,	12 / 2*	Rumilly to Nine wood		
2nd Bde.	B	,,	6th	16th	24 / 4*	Blue	Marcoing.	
,,	,,	,,	,,	,,	12 / 2*	Brown		
,,	H	,,	,,	71st	24 / 4*	Blue	Nine wood	
,,	,,	,,	,,	,,	12 / 2*	Brown		
1st Bde.	D	III	51st	152nd	42	Blue	Fontaine Bourlon wood, Bapaume-Cambrai road	1st Bde. used all tanks in mechanical reserve.
,,	E	,,	,,	153rd	28	Flesquières		
,,	,,	,,	62nd	186th	14	Brown	Bourlon village, Graincourt	
,,	G	,,	,,	185th	42	Havrincourt		

* In mechanical reserve, to replace breakdowns.

ten minutes before zero hour, the tanks, which had deployed on a line some 1,000 yards from the enemy's outpost trenches, began to move forward, infantry in section columns advancing slowly behind them. Ten minutes later, at 6.20 a.m., zero hour, the 1,000 British guns opened fire, the barrage coming down with a terrific crash about 200 yards in front of the tanks which were now proceeding slowly across "No Man's Land," led by Brigadier-General H. J. Elles, the Commander of the Tank Corps, who flew the Tank Corps colours from his tank and who on the evening before the battle had issued the following inspiring Special Order to his men:

Special Order No. 6

1. Tomorrow the Tank Corps will have the chance for which it has been waiting for many months—to operate on good going in the van of the battle.

2. All that hard work and ingenuity can achieve has been done in the way of preparation.

3. It remains for unit commanders and for tank crews to complete

BATTLE OF CAMBRAI
November 20th 1917.

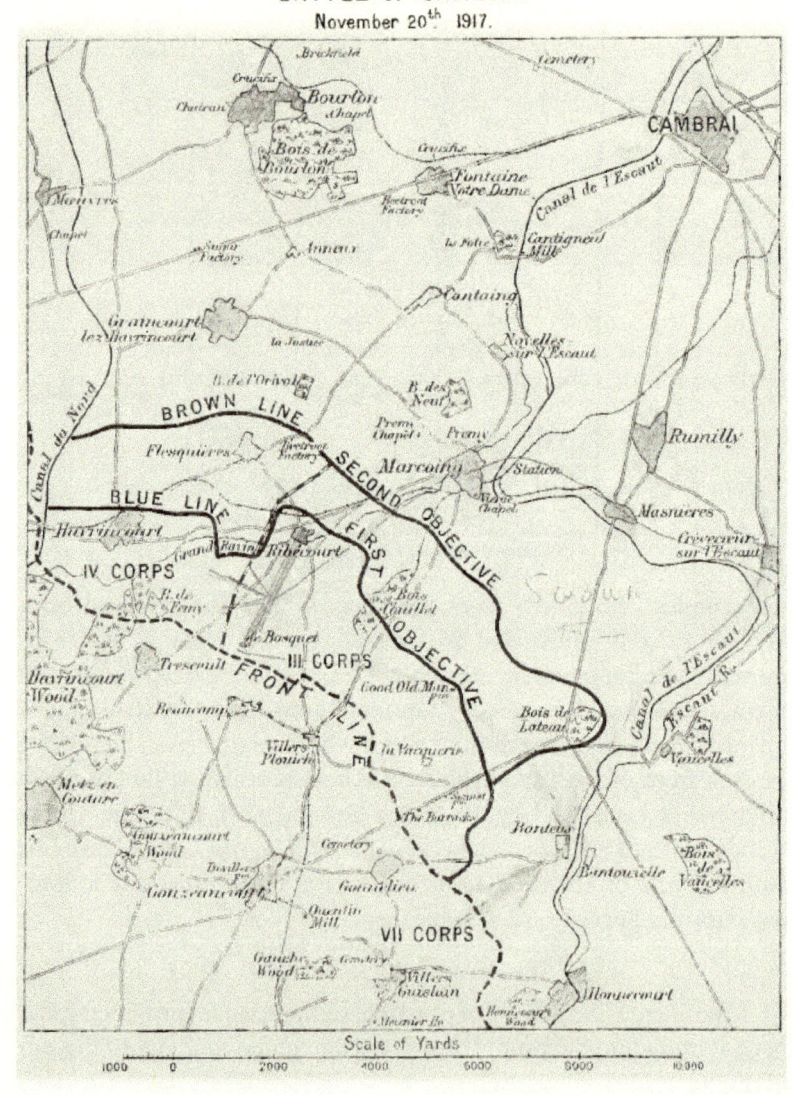

the work by judgment and pluck in the battle itself.

4. In the light of past experience I leave the good name of the Corps with great confidence in their hands.

5. I propose leading the attack of the centre division.

Hugh Elles,
B.G. Commanding Tank Corps.

November 19, 1917.

(The statement made in the daily press that General Elles' order ran—"England expects every tank to do its damnedest," was a pure journalistic invention and one in very bad taste.)

The attack was a stupendous success; as the tanks moved forward with the infantry following close behind, the enemy completely lost his balance and those who did not fly panic-stricken from the field surrendered with little or no resistance. Only at the tactical points was opposition met with. At Lateau wood on the right of the attack heavy fighting took place, including a duel between a tank and a 5·9 in. howitzer. Turning on the tank the howitzer fired, shattering and tearing off most of the right-hand sponson of the approaching machine, but fortunately not injuring its vitals; before the gunners could reload the tank was upon them and in a few seconds the great gun was crushed in a jumbled mass amongst the brushwood surrounding it. A little to the west of this wood the tanks of F Battalion, which had topped the ridge, were speeding down on Masnières.

One approached the bridge, the key to the Rumilly-Seranvillers position, upon the capture of which so much depended. On arriving at the bridge, it was found that the enemy had already blown it up, nevertheless the tank attempted to cross it; creeping down the broken girders it entered the water but failed to climb the opposite side. Other tanks arriving and not being able to cross assisted the infantry in doing so by opening a heavy covering fire. Westwards again La Vacquerie was stormed and Marcoing was occupied. This latter village had been carefully studied beforehand and a definite scheme worked out as to where tanks should proceed after entering it. Difficult though this operation was, each position was taken up and the German engineers shot just as they were connecting up the demolition charges on the main bridge to the electric batteries.

In the Grand Ravin, which runs from Havrincourt to Marcoing, all was panic, and from Ribecourt northwards the flight of the German soldiers could be traced by the equipment they had cast off in order to speed their withdrawal. Nine wood (Bois des Neuf) was stormed, and Premy Chapel occupied. At the village of Flesquières the 51st Division, which had devised an attack formation of its own, was held up; it appears that the tanks out-distanced the infantry or that the tactics adopted did not permit of the infantry keeping close enough up to the tanks. As the tanks topped the crest, they came under direct artillery fire at short range and suffered heavy casualties. This loss would have mattered little had the infantry been close up, but, being some distance off, directly the tanks were knocked out the German machine gunners, ensconced amongst the ruins of the houses, came to life and delayed their advance until nightfall; thus, Flesquières was not actually occupied until November 21.

In the village of Havrincourt some stiff fighting took place. All objectives were, however, rapidly captured, and the 62nd Division had the honour of occupying Graincourt before nightfall, thus effecting the deepest penetration attained during the attack on this day. From Graincourt several tanks pushed on towards Bourlon wood and the Cambrai road, but by this time the infantry were too exhausted to make good any further ground gained.

Meanwhile No. 3 Company of A Battalion had assisted the 29th Division on the Premy Chapel-Rumilly line, one section of tanks working towards Masnières and another co-operating with the infantry in the attack on Marcoing and the high ground beyond. The third section attacked Nine wood, destroying many machine guns there and at the village of Noyelles, which was then occupied by our infantry.

Whilst these operations were in progress the supply tanks had moved forward to their "rendezvous," the wireless-signal tanks had taken up their allotted position, one sending back the information of the capture of Marcoing within ten minutes of our infantry entering this village; and the wire-pullers cleared three broad tracks of all wire so that the cavalry could move forward. This they did, and they assembled in the Grand Ravin and in the area adjoining the village of Masnières.

By 4 p.m. on November 20, one of the most astonishing battles in all history had been won and, as far as the Tank Corps was concerned, tactically finished, for, no reserves existing, it was not possible to do more than rally the now very weary and exhausted crews, select the fittest, and patch up composite companies to continue the attack on the morrow. This was done, and on the 21st the 1st Brigade supported the 62nd Division with twenty-five tanks in its attack on Anneux and Bourlon wood and the 2nd Brigade sent twenty-four machines against Cantaing and Fontaine-Notre-Dame, both of which villages were captured.

November 21 saw, generally speaking, the end of any co-operative action between tanks and infantry; henceforth, new infantry being employed, loss of touch and action between them and the tanks constantly resulted. Nevertheless, on the 23rd a brilliant attack was executed by the 40th Division, assisted by thirty-four tanks of the 1st Brigade; this resulted in the capture of Bourlon wood. The tanks then pressed on towards the village; the infantry, however, who had suffered severe casualties in the capture of the wood, were not strong enough to secure a firm footing in it.

This day also saw desperate fighting in the village of Fontaine-Notre-Dame. Twenty-three tanks entered this village in advance of our own infantry; there they met with severe resistance, the enemy retiring to the top stories of the houses and raining bombs and bullets down on the roofs of our machines. Our infantry, who were very exhausted, were unable to make good the ground gained, consequently, all tanks which were able to do so withdrew under cover of darkness at about 7 p.m.

On November 25 and 27 further attacks were made by tanks and infantry on Bourlon and Fontaine-Notre-Dame with varying success, but eventually both these villages remained in the hands of the enemy. So ends the first phase of the Battle of Cambrai.

During the attacks which had taken place since November 21, tank units had become terribly disorganised, and by the 27th had been reduced to such a state of exhaustion that it was determined to withdraw the 1st and 2nd Brigades. This withdrawal was nearing completion when the great German counter-attack was launched early on the morning of November 30.

To appreciate this attack, it must be remembered that at this time the IIIrd and IVth Corps were occupying a very pronounced salient, and that all fighting had, during the last few days, concentrated in the Bourlon area and had undoubtedly drawn our attention away from our right flank east of Gouzeaucourt. The plan of General von der Marwitz, the German Army Commander, was a bold one, it was none other than to capture the entire IIIrd and IVth British Corps by pinching off the salient by a dual attack, his right wing operating from Bourlon southwards and his left from Honnecourt westwards, the two attacks converging on Trescault. Between these two wings a holding attack was to be made from Masnières to La Folie wood.

The attack was launched shortly after daylight on November 30, and failed completely on the right against Bourlon wood. Here the enemy was caught by our artillery and machine guns and mown down by hundreds. On the left, however, the attack succeeded: firstly, it came as a surprise; secondly, the Germans heralded their assault by lines of low-flying aeroplanes which caused our men to keep well down in their trenches and so lose observation. Under the protection of this aeroplane barrage and a very heavy mortar bombardment the German infantry advanced and speedily captured Villers Guislain and Gouzeaucourt.

At 9.55 a.m. a telephone message from the IIIrd Corps warned the 2nd Brigade of the attack, but, in spite of the fact that many of the machines were in a non-fighting condition, by 12.40 p.m. twenty-two tanks of B Battalion moved off towards Gouzeaucourt, rapidly followed by fourteen of A Battalion. Meanwhile the Guards Division recaptured Gouzeaucourt, so, when the tanks arrived, they were pushed out as a screen to cover the defence of this village. By 2 p.m. twenty tanks of H Battalion were ready, these moved up in support.

Early on the morning of December 1, in conjunction with the Guards Division and 4th and 5th Cavalry Divisions, the 2nd Brigade delivered a counter-attack against Villers Guislain and Gauche wood. The western edge of the wood was cleared of the enemy; the tanks then proceeded through the wood, where very heavy fighting took place. From the reports received as to the large number of dead in the wood and the numerous machine guns found in position, it is clear that the enemy had intended to hold it at all costs. Once the wood was cleared the tanks proceeded on to Villers Guislain, but being subjected to direct gun fire eventually withdrew.

The counter-attack carried out by the 2nd Brigade greatly assisted in restoring a very dangerous situation; it was a bold measure well executed, all ranks behaving with the greatest courage and determination under difficult and unexpected circumstances and amidst the greatest confusion caused by the success of the German attack; every tank crew of every movable machine had but one thought, namely, to move eastward and attack the enemy. This they did, and it is a remarkable fact that, though at 8 a.m. on November 30 not one machine of the Brigade was in a fit state or fully equipped for action, by 6 a.m. on the following day no fewer than seventy-three tanks had been launched against the enemy with decisive effect.

Thus, ended the first great tank battle in the whole history of warfare, and, whatever may be the future historian's dictum as to its value, it must ever rank as one of the most remarkable battles ever fought. On November 20, from a base of some 13,000 yards in width, a penetration of no less than 10,000 yards was effected in twelve hours; at the Third Battle of Ypres a similar penetration took three months. Eight thousand prisoners and 100 guns were captured, and these prisoners alone were nearly double the casualties suffered by the IIIrd and IVth Corps during the first day of the battle. It is an interesting point to remember that in this battle the attacking infantry were assisted by 690 officers and 3,500 other ranks of the Tank Corps, a little over 4,000 men, or the strength of a strong infantry brigade, and that these men replaced artillery wire-cutting and rendered unnecessary the old preliminary bombardment.

More than this, by keeping close to the infantry they effected a much higher co-operation than had ever before been attainable with artillery. When on November 21 the bells of London pealed forth in celebration of the victory of Cambrai, consciously or unconsciously to their listeners they tolled out an old tactics and rang in a new—Cambrai had become the Valmy of a new epoch in war, the epoch of the mechanical engineer.

CHAPTER 20
An Infantry Appreciation of Tanks

During the many battles and engagements in which the Tank Corps took part many appreciative special orders and letters were received from the Higher Commanders under whose orders the Corps worked. These kindly words, always appreciated, are apt sometimes to be regarded as the inevitable "good chits" which courtesy demands should be addressed to good, indifferent, and bad alike after an operation has been successfully completed. Unsolicited testimonials, and especially such as are not meant for the eyes of those praised, when they do, by chance, come under these eyes, are regarded as more than mere "pats on the back," especially when they come from those who have fought alongside the commended.

The following letter was written by an infantry officer who took part in the Battle of Cambrai, and addressed to a personal friend neither in nor connected with the Tank Corps, who, months later on, showed it to one who was. Not only did this letter come as a pleasant and gratifying surprise to all ranks of the Corps, for it was published as a "Battle Note," but it shows such an exceptionally clear insight into the value and possibilities of the tanks that, even for this reason alone, it is worth publishing. Who the writer was the Tank Corps never knew, but his sound judgment and kindly appreciation stimulated amongst his readers that high form of personal and collective pride which to soldiers is known as *"esprit de corps."*

★★★★★★

"Battle Notes" were issued from time to time by Tank Corps Headquarters to all tank crews. Their object was to stimulate *"esprit de corps* and moral." They were human documents for the most part, referring not only to the tank but also to other arms.

★★★★★★

It is on human documents such as these, rather than on orders and instructions, that the moral of an individual or a unit grows strong, and by growing strong places the entire army one step nearer victory.

The letter reads as follows:

I will first give you the opinion of one of my colonels. In three years of fighting on this front, I have met no battalion commander to equal him in power of leadership, rapidity of decision in an emergency, and personal magnetism. I have met no man who would judge so justly what an infantry soldier *can* and *cannot* do.

He considers the tanks *invaluable* if properly handled, either for the attack or in defence—but he realises, as I think we all do, that until Cambrai, the tactical knowledge shown in their employment was of the meanest order.

One other valuable opinion I've obtained. We have now with the battalion a subaltern, a man of about thirty—a very good soldier, a resolute, determined kind of fellow who has seen a good deal of fighting. He commanded a platoon in our —th Battalion in the big tank attack at Cambrai and was in the first wave of the attack throughout. He tells me that the tanks covering the advance of his battalion functioning under ideal weather and ground conditions, were handled with marked skill and enterprise in the capture of the first two objectives covering an advance of about 3,500 yards.

The morale effect of the support given by tanks on the attacking infantry is *very great*. He says his men felt the utmost confidence in the tanks and were prepared to follow them anywhere. The effect of the advancing line of tanks on the enemy infantry was extraordinary. They made no attempt whatever to hold their trenches, and either bolted in mad panic or, abandoning their arms, rushed forward with hands uplifted to surrender. As long as the advance of the tanks continued, *i.e.*, over the enemy trench system to a depth of from two to three miles, the total casualties incurred by our —th Battalion (attacking in the first wave) were four killed and five wounded, all by shell fire.

After the fall of the second objective, the advance ceased for some unexplained reason—(they were told some hitch about Flesquières)—the attack seemed to lose purpose and direction. Tanks on the flanks began coming back. The battalion was ordered to attack five different objectives, and before the necessary plans could be communicated to subordinate commanders, orders were received cancelling the previous instructions.

In a word, chaos prevailed. The afore-mentioned subaltern cannot speak too highly of the work of the tank commanders—nothing could exceed their daring and enterprise. He says he is absolutely convinced that infantry, unsupported by artillery, are absolutely powerless against tanks and that no belt of wire can be built through which they cannot break an admirable passage for infantry.

Lastly, he makes no secret of the fact that it would demand the utmost exercise of his determination and resolution to stand fast and hold his ground in the face of an attack by enemy tanks, carried out on the same scale as ours. I may add that he is a big upstanding fellow, a fine athlete, and afraid of nothing on two legs.

I give you his opinion at some length, because they are the *ipsissima verba* of a man qualified to speak from personal practical experience. Personally, I believe the tanks may yet play the biggest role in the war, if only the Higher Command will employ them in situations where common-sense and past experience alike demand their use. Two days before the Hun attacked us at Bourlon wood we lost three officers and some seventy gallant fellows trying to mop up a couple of enemy M.G. nests—a bit of work a couple of tanks could have done *with certainty* without the loss of a man.

In the situation described after the capture of the second objective, why should there not have been a responsible staff officer—G.S.O.1 say—right forward in a tank to size up the situation and seize opportunity, the very essence of which is rapid decision? In the early days of the war, forgetful of the lessons of South Africa, we put our senior officers in the forefront of the battle—of late, the pendulum has swung the other way—surely the employment of a tank for the purpose outlined would enable us now to strike the happy mean?

In defence, as a mobile 'pill-box,' the possibilities of the tank are great—any man who has led infantry 'over the top' knows the demoralising and disorganising effect of the 'surprise packet' machine-gun nest—what more admirable type of nest can be devised? Continually changing position, hidden from enemy aircraft by smoke and the dust of battle, offering no target for aimed artillery fire.

Half the casualties we suffer in heavy fighting after the initial attack come from the carrying parties winding slowly in and out through barrage fire, bringing up ammunition to the infantry, the Lewis and Vickers guns—all this could be done much more rapidly, surely, with a minimum of loss, by tanks—for the future the tank should relieve the artillery of all responsibility as regards wire-cutting. You *know* you can cross a belt of wire over which a tank has passed, you *hope* you can pass through a wire belt on which the artillery has played for a couple of days. As a business proposition a tank at £5,000 will cut more wire in one journey, even assuming it does nothing else, than 2,000 shells at £5 each, blazing away for a day—add the wear on the life of the gun.

In attack, one of the most difficult problems of the infantry is to get the Stokes guns far enough forward, with sufficient ammunition, to come into action against machine guns or strong points holding up the advance unexpectedly—all this could be done by means of a tank with ease—whilst not only could the small Stokes gun with a range of 500–600 yards be brought forward, but also the 6 in. Stokes with a range of 1,200–1,600 yards by the same means, and be brought into action firing from the tank.

The tank has only one enemy to fear—the high-velocity tank gun, firing aimed shots from forward positions. I believe this danger can be minimised by means of escort aeroplanes attached during an action to every tank, and provided with smoke bombs to blind the gun position, if unable to silence the gun by machine-gun fire or by means of ordinary bombs heavily charged.

I have tried to outline some of the more obvious uses for which the tank is so admirably suitable. There is a well of this information yet untapped, not in staff offices, but in the minds of the platoon and company commanders who have fought in the first waves of the attack with the tank, who have seen the difficulties it has to overcome and how it has met them or failed, and why. Nothing has yet been produced in this war to equal the tank for doing by *machinery* what has hitherto been done by *men*; nothing so well fitted to economise our man-power and reduce the appalling wastage which has hitherto characterised our efforts in attack, and with gain instead of loss in efficiency. We want thousands of tanks, both light and heavy, ranging from two miles to eight miles per hour, armed with M.G.s, armed with Stokes guns, unarmed and fast-travelling for transport of gun teams to emergency tactical positions, and lastly, a staff of trained minds to define the tactics of the tank—to refute criticism based on ignorance, to collect, classify and investigate all available information and suggestions, so that like an aeroplane—every 'new edition' of the tank is an improvement on the last.

I have written at some length, but the subject is big and attractive enough to be my excuse.

Chapter 21
The Tank Corps Training Centre

Early in February 1916 a Conference was held at the War Office, to decide as to the training of the personnel for the tank units it was now decided to raise. At this Conference, Lieutenant-Colonel Swinton and Lieutenant-Colonel R. W. Bradley, D.S.O., were ordered to be present.

At this time, Lieutenant-Colonel Bradley was Commandant of the Motor Machine Gun Training Centre at Bisley, and was in the position to select suitable men for the new arm.

The number of men required for the first 150 tanks was estimated at 1,500, or ten men for each machine, and 150 junior officers. This personnel was obtained as follows: 600 men were transferred from the reserves in training at the Motor Machine Gun Training Centre and 900 were obtained by special enlistment. Thirty officers were transferred from the Motor Machine Gun Section, fifteen were detailed by G.H.Q., France, and the remainder were obtained by calling for volunteers from units in England and by special selection from Cadet units.

For purposes of secrecy the new formation was "tacked on" to the Machine Gun Corps and was christened with the terrific name of: "Special Armoured Car Section Motor Machine Gun Section." A month later it became known as the Heavy Section Machine Gun Corps.

The recruiting was very successful, and this was largely due to the untiring energy of Mr. Geoffrey Smith, editor of *The Motor Cycle*, who spared neither time, trouble, nor money in getting the men.

Towards the end of March, the training camp was moved from Bisley to Bullhouse farm, and at this camp all elementary training was carried out, recruits being taught drill, the ways of military life, and the Vickers and Hotchkiss machine guns as well as the Hotchkiss 6-pounder.

The first establishment issued by the War Office provided for 10 companies of 10 tanks each, but within a fortnight this was changed to 15 companies of 10 tanks each, the companies being grouped in 3 battalions. A little later this organisation, at the request of G.H.Q., France, was again changed to one of 6 companies of 25 tanks each.

With a further view of ensuring secrecy it was arranged by Lieutenant-Colonel Swinton that no tanks should be sent to Bisley, but that a training ground, far removed from prying eyes, should be secured. Steps were at once taken to find such a ground, and eventually Thetford in Norfolk was visited and Lord Iveagh's estate at Elveden selected. The necessary training ground here was taken over and was known as "the Elveden Explosives Area"; and round it at 200 yards intervals were posted groups of sentries of the Royal Defence Corps.

During the early part of April, a certain amount of grumbling on the part of specially enlisted men occurred at Bisley. They had been induced to join an Armoured Car Service, and for six weeks they had not even seen the wheel of a car. They were asked to exercise a little patience and were promised a surprise. At Elveden the surprise was revealed to them, and when they had got over their astonishment on seeing the first Mark I tank approach them they set down to work with a will, which, it is an honour to record, was never abandoned by all ranks of the Tank Corps from this day on to the conclusion of the war.

The camp at Elveden was placed just outside the "Explosives Area" and no one was allowed to enter the area without a special permit.

Companies, before they proceeded overseas, however, spent their last three weeks within the area. As soon as this necessary ground had been taken over, three pioneer battalions were brought to Elveden Camp and a complete series of trenches was dug on a front of a mile and a quarter, and to the depth of two miles. The plan of this work was laid out by Major Tandy and Captain Martel, both R.E. officers.

Unfortunately, on account of delay in delivery of tanks, constructional defects and repeated requests from G.H.Q., France, that all available tanks and crews should be sent out to France for the September operations on the Somme, little use was made of these trenches, for tactical training. Machine-gun firing from tanks with ball ammunition was, however, freely carried out, and also 6-pounder practice which, unfortunately, was much hampered by danger restrictions.

The tank drivers were all drawn from the A.S.C., and the 711th Company A.S.C. was formed to include these men, the workshops, and the M.T. personnel of the Heavy Section. The officer in command of this company and in charge of all mechanical instruction and driving was Major H. Knothe, D.S.O., M.C.

By the end of May the last company had completed its training at Bisley and had moved to Elveden; the headquarters, having some time prior to this, moved to this place and established itself in the stables of Lord Iveagh's mansion and in the new alms-houses in Elveden village.

By the beginning of July training was sufficiently advanced to give the first tank demonstration ever held. Twenty tanks took part in it and advanced in line followed by infantry against a section of the instructional trench system. The demonstration was a great success and many notable persons witnessed it, including Mr. Lloyd George and Sir William Robertson.

This demonstration was shortly afterwards followed by a second at which the king attended. His Majesty was most anxious that his projected visit should be kept secret, but as it was nevertheless necessary to make certain preparations it was given out at the camp that a very distinguished Russian general was about to visit the tanks. The identity of the Russian general was, however, discovered by the bulk of the men before the demonstration was concluded, much to their pleasure and amusement.

At the beginning of August Lieutenant-Colonel Brough, C.M.G., visited G.H.Q., France, to ascertain the tactics it was proposed to employ as regards tanks. Unfortunately, his visit was fruitless, for no ideas apparently existed on the subject. Shortly after his return instructions were received to dispatch the tank companies to France, and to decide on this a conference was held at which the following officers were present: Major-General Butler, Brigadier-General Burnett-Stuart, both from G.H.Q., France; Lieutenant-Colonel Swinton, Lieutenant-Colonel Bradley, and Lieutenant-Colonel Brough. At this Conference it was decided to mobilise the companies at Elveden and to dispatch them overseas by half companies. The first to leave was C Company and the second D Company, which, respectively, were under the commands of Majors Holdford-Walker and Summers.

Towards the end of August Colonel Swinton was instructed to send over to France a staff officer, but as the establishments only allowed of a commander and one staff captain, and as the latter was a very junior officer, Lieutenant-Colonel Brough was sent over. Shortly after his arrival he found it necessary to wire home for motorcars, clerks, etc., for he had been ordered to take over command of the units in France. Captain Kingdon was thereupon sent out to assist him, and two clerks and a motorcar were also dispatched. By these means were extemporised an advanced headquarters, the original headquarters of the Heavy Branch remaining in England and never proceeding overseas.

In October 1916, as already related in Chapter 6, Bovington Camp, Wool, was selected as the new training centre. Here E, F, G, H, and I Battalions were raised and trained during 1916–1917, and J, K, L, M, N, O, P, Q, and R during 1917–1918, the last battalion, the 18th, sailing for France in September 1918.

In 1917, to cope with the steadily increasing number of tank units of all descriptions, Worgret Camp, Wareham, and Lulworth were taken over, the Depot Reserve Unit being established at the former and the Gunnery Camp at the latter place.

The first schools to be formed were the Tank Drivers' School, the 6-Pounder School, and the Lewis Gun School, but by degrees, as the Tank Corps grew, these developed until at the close of the war the following schools had been established:

Tank Drivers and Maintenance School.
Tank Gunnery School (6-Pounder and Machine Gun).
Tank Reconnaissance School.
Tank Signal and Pigeon School.
Camouflage School.
Revolver School.
Gas School.

Tank Compass School.

In July 1918 the preparations set on foot to double the Tank Corps for 1919 threw a great deal of work on to the Training Centre. Thirteen British, three Canadian, and one New Zealand Battalion were to be raised, as well as a number of subsidiary units. In August, in spite of shortage of infantry reinforcements, an allotment, given precedence over all the other arms, of 4,500 men, was made to the Tank Corps Training Centre, so that the raising of the above new units might forthwith begin; besides this, nearly half a million pounds worth of buildings were sanctioned without estimates being called for, so important was it now considered that not a day should be lost in the Tank Corps preparations for 1919.

By the date of the armistice about half the building programme was finished, and eight British and one Canadian battalion had been raised.

The following is a summary of the total number of tank units and reinforcements raised and trained at the Training Centre between November 1916 and November 1918.

British Tank Battalions	22 (5th to 26th).
Canadian Tank Battalions	1 (1st Canadian Tank Battalion).
American Tank Battalions	3 (301st, 302nd, and 303rd).
Gun Carrier Companies	2 (1st and 2nd).
Tank Supply Companies	5 (1st to 5th).
Tank Advanced Workshops	2 (Nos. 4 and 5).
Tank Salvage Companies	1 (No. 3).
American Tank Salvage Companies	2 (306th and 317th).
Various Headquarters	3
Total Tank Units raised	41

The whole of the above units, with the exception of eight British and one Canadian Battalion, were sent out to France prior to the armistice.

In all, some 21,000 officers and men passed through the Training Centre, 14,000 in formed units, and 7,000 as reinforcements; besides these, 950 cadets were trained. In October 1918 the Training Centre, which from one camp at Bovington had grown to include Worgret, Lulworth, and Swanage Camps, had on its strength in all ranks and service approximately 16,000 men.

The time required wherein to raise and train a new Tank Battalion averaged four months. The system of instruction adopted from November 1916 onwards was to start with a very thorough individual training and then to pass the recruits through the various schools, leaving collective and tactical training to be carried out in France.

Recreational training played an important part in the above instruction, and the Training Centre gained a considerable reputation in the Southern Command for efficiency in sports and games.

In the expansion which commenced on September 1, 1918, 30 *per cent.* of the personnel for each new unit was sent to the Training Centre from the trained Tank Corps personnel in France, and this trained personnel, together with the increased numbers of training tanks and other improved facilities, would have gone far to effect a more efficient and rapid training of the units, before their departure overseas, than heretofore.

Besides raising and training new units and reinforcements the Tank Corps Training Centre was intimately connected with much of the experimental work, armament design, and the fittings of all types of tanks from the introduction of the Mark V and Medium "A" tanks onwards. The following were the main improvements initiated.

The adaptation of the Hotchkiss machine gun to the tank.

The invention of the Palmer machine-gun battle-sight.

The invention of fire-control instruments.

During the spring and summer of 1917 various experiments were carried out at Wool to arrive at the best method of demolishing and removing wire entanglements. Eventually grapnels were decided upon and were used with great success in November at the Battle of Cambrai.

The use of cloud smoke from tanks was also originated at the Training Centre, and with the aid of an invention of the late Commander Brock was eventually adopted for all tanks, and was used on several occasions with effect during the summer and autumn operations of 1918.

For purposes of general interest and education as well as for the conversion of the mechanical heathen, a considerable number of demonstrations, showing the power of tanks and their co-operation with infantry, were given to officers of the War Office, Commands and Schools throughout 1918. On October 25 this year.

His Majesty the King visited Wool to witness one of these, and paid the Tank Corps Training Centre the great honour of inspecting the various battalions, and welcomed many of the men of the British and American units assembled by walking amongst them and conversing freely with them.

CHAPTER 22

The Tank Supply Companies

Tanks, like every other arm of the army, require a highly organised supply service, and being cross-country machines they must be served by machines of similar powers of locomotion. This was probably realised before tanks were originally dispatched to France in 1916, but, during the battles of the Somme, Ancre, and Arras, it was not possible to organise any system of cross-country supply on account of every machine being required for either fighting or training purposes. In February 1917 the first organisation for cross-country supply was formulated. This consisted in allotting two supply tanks to each company, but the delay in the arrival of Mark IV machines prevented this organisation taking form until May 1917.

Supply tanks were first employed at the Battle of Messines, the Mark I tanks, which had now been discarded as fighting machines, being used for this purpose. These machines were fitted with large soft steel sponsons made at the Tank Corps Central Workshops. During this battle they were not much used owing to the limited scope of the operations.

Between June 1917 and the opening of the Third Battle of Ypres each tank battalion received six supply tanks, two for each company, but during this battle they did not prove a success on account of the appalling conditions of the ground, the sponsons continually becoming wedged in the shell-holes, which necessitated much digging out to relieve them.

Just prior to the opening of this battle the first of the gun-carrier tanks arrived in France, and was dispatched to Flanders and attached to the XVIIIth Corps for experiment. Later on, others followed, until by the end of the year forty-four of these machines had been received.

The idea of the gun-carrier was that of mechanical artillery, the machine being really a mechanical gun mounting capable of carrying a 60-pounder or 6 in. howitzer. Its total shell capacity without the gun was 200 6 in. shells, weighing approximately 10 tons.

Considering the difficulties of the ground very good work was done by the gun-carriers during the Ypres operations, several hundreds of tons of ammunition being carried forward as well as a few 60-pounders.

In September a new method of supply was experimented with; this consisted in towing behind any type of tank three sledges connected with the roof of the machine by a cable. At the Battle of Cambrai this method proved a great success, and not only were tank supplies hauled forward but also telegraph cable and bridging material.

During the autumn and winter of 1917 much careful thought had been devoted both in France and England to the question not only of tank supply but of being able to carry forward infantry, particularly machine-gunners, in armoured carrier tanks; the result of this was the design of a large carrier tank known as the Mark IX and the raising of a new unit known as an "Infantry Carrier Company."

These carrier units were first formed on February 1, 1918. The first two companies consisted mostly of Royal Engineer personnel, and the next three of infantry. The standard of the personnel was very good, about 60 *per cent.* having already seen service overseas. The 1st and 2nd Companies proceeded to France about the middle of May, the remaining three arriving in June and July.

The organisation of each of these companies was as follows:

A company headquarters and four sections, each section consisting of six Mark IV supply tanks, or tenders, as they were sometimes called. The object of these companies was laid down in Tank Corps Standing Orders as follows:

> The Tank Supply Company is a unit of Brigade Troops for the carriage of supplies, from the point where wheeled vehicles cease, to battalions. The responsibility for maintaining battalion supplies rests with Brigade Headquarters. The duties of the Brigade Supply Officer will in no way be shared by the O.C. Tank Supply Company. The Tank Supply Company will be used as a mobile reserve of supplies under the immediate control of Brigade Headquarters.

These supply companies were never used for carrying forward infantry, as the Mark IX tank did not materialise until October 1918; they proved, however, of the greatest use during all the tank operations of the last year of the war.

During June the two Gun Carrier Companies were definitely converted into Supply Companies and were attached to the 3rd and 5th Brigades. At the Battle of Hamel, on July 4, four of these machines did excellent work, carrying forward between twenty and twenty-five tons of R.E. material and dumping this a few hundred yards behind the final objective within half an hour of this objective being captured. These machines were driven by four officers and sixteen men, and had the material they transported been brought up by carrier parties at least 1,200 men would have been required; in man-power alone these four machines thus saved 1,184 soldiers, or approximately two infantry battalions at battle strength.

On arrival in France the 1st and 2nd Supply Companies were posted to the 1st and 4th Tank Brigades, and the 3rd, 4th, and 5th were sent to Blingel Camp, in the Bermicourt area, where good facilities existed for tank driving and maintenance. At about the end of July the 3rd and 5th Companies were equipped with Mark IV supply tanks, and female Mark IV machines fitted with a sledge equipment.

At the beginning of August, the distribution of the various supply units was as follows:

No. 1 Gun Carrier Company	5th Tank Brigade
No. 2 Gun Carrier Company	3rd Tank Brigade
No. 1 Tank Supply Company	1st Tank Brigade
No. 2 Tank Supply Company	4th Tank Brigade
No. 3 Tank Supply Company	Blingel Camp
No. 4 Tank Supply Company	2nd Tank Brigade
No. 5 Tank Supply Company	Blingel Camp

All these companies, less No. 1 Tank Supply Company and No. 2 Gun Carrier Company, took part in the Battle of Amiens.

No. 1 Gun Carrier Company suffered an unfortunate experience on August 7. It had moved forward to an orchard on the western side of Villers Bretonneux, each of its machines being loaded up with explosives of various kinds. A shell fired from a German battery in the vicinity of Chipilly set fire to one of the camouflage nets, and the result of this was that though six out of the twenty-two machines got away the remaining sixteen were blown up, the explosion being terrific.

The 3rd Tank Supply Company was allotted to the Canadian Corps to carry forward infantry supplies such as grenades, S.A.A., and drinking water. The female Mark IV. tanks equipped with sledges were attached to the Canadian Engineers for the purpose of bringing forward material in order to repair the bridges over the Luce River. Owing to weak cables this operation proved a failure, most of these machines breaking down before they had covered a mile.

The policy which was first adopted of attaching a section of six supply tanks to each battalion did not work well, the Company Headquarters was usually left in the air, and soon lost touch with its sections. In order to remedy this defect from August 9 onwards company commanders were instructed to establish "report centres" well in advance of the battlefield. These centres were "baited" by sending the mails there; to obtain news from home it was consequently necessary for section commanders to send runners in to fetch them; by this means touch with the Company Headquarters was automatically maintained.

★★★★★★

From this it must not be deduced that the officers and men of the Tank Corps would not obey orders, but that the officer in command of the Supply Companies was a student of human nature. Why order when a simple act like this will do the ordering?

★★★★★★

In the battles north of the Somme, commencing on August 21, much useful work was carried out, the tank-drivers having by now become thoroughly expert in driving and maintenance. The sections were now properly brigaded, each company being looked upon as a unit and not as a mere headquarters for four separate units. Proper telephonic communication was now established between the sections and the company, and consequently much time was saved not only within the company itself but by the various units it was supplying.

During all the battles onwards from August 8 to the capture of Landrecies the work carried out by the Tank Supply Companies and the Gun Carrier Companies was not only useful but of great importance, as in many places the roads were too bad for mechanical transport.

When they were not required to bring forward tank supplies, they were engaged in carrying every sort of ammunition and engineer stores, especially through zones which were harassed by machine-gun fire and in which, had infantry carrying parties been used, many lives would have been lost.

When the possibilities of these companies became realised, infantry commanders were continually asking for their assistance, preference being given to the gun-carriers on account of their greater capacity for light stores.

The Gun Carrier Companies, besides doing excellent work as infantry supply companies, kept both field and heavy artillery well supplied. No. 2 Gun Carrier Company carried out some very successful heavy sniping by carrying forward a 6 in. howitzer, and by moving it from place to place during the night it both harassed and puzzled the enemy. Besides this, several successful gas attacks were carried out with the aid of the gun-carriers, which transported the projectors and bombs to positions over country which wheeled transport could not have negotiated. By using these machines, it was possible to get in three or more "shoots" in one night and to retire out of the danger zone before dawn.

If in the days of the great Napoleon, when a soldier went into action with frequently less than twenty balls in his pouch and a couple of spare flints, an army "crawled on its stomach," how much more does it crawl today! When the lessons of the war are sorted and tabulated in order of importance, very near the top, if not at the top itself, will be found that of "road capacity," in other words, that victory rests with the side which can maintain the broadest communications. To widen existing roads directly by enlarging them or to construct new roads are both works of great labour; they absorb not only time and men but also transport of every kind, especially in a country like north-eastern France, where suitable stone for road-metalling is practically non-existent. To do so indirectly is best accomplished by a cross-country tractor, that is, by a machine which can move on or off a road. With such a machine, roads can be indefinitely widened; paradoxically they cease to exist, for they are no longer necessary.

The tank is, first of all, a cross-country tractor, and it is curious that none of the contending nations appear to have appreciated this until well towards the end of the war, in spite of the fact that the reason for the general slowness of the advances which followed any initial success was nearly always due to inadequacy of supply.

By the end of March 1918, the German attack "petered out" for want of supplies; by the end of May it again did likewise for a similar reason. Had the Germans possessed on March 21 and May 27 5,000 to 6,000 efficient cross-country tractors, each of which could have carried five tons of supplies, all the hosts of brave men, which the United States of America could have poured into France, could not have prevented a separation of the British and French Armies from being effected.

Had such a separation taken place it is impossible to say what the result might not have been; but what is possible to say is that had the Germans "scrapped" half their guns and replaced them by cross-country tractors they would have gone nearer winning the war than they did.

CHAPTER 23
The Second Battle of the Somme

With the close of the Battle of Cambrai the British Army abandoned the offensive, which had been initiated on April 9, and a period of passive defence was developed. At this time all three Tank Brigades had assembled at or near Bray-sur-Somme, where extensive hutments existed and where the old devastated area offered excellent facilities for training. Towards the end of December a request was made by the Tank Corps to establish at Bray a large tank and infantry school, so that co-operation between these two arms might be secured; further, as artillery ranges were near at hand it was felt that a complete tactical unity of action between tanks, infantry, artillery, and aeroplanes could now be established: besides this, Bray formed an excellent strategical centre to the Somme area should the Germans at any time launch an attack between the Oise and the Scarpe.

Early in January 1918 orders were, however, received that in place of remaining assembled at one spot the Tank Corps was to form a defensive cordon stretching from about Roisel to a little south of Bethune—a frontage of some sixty miles. In February this line was taken up, tank units being distributed as follows:

Fifth Army, 4th Tank Brigade	1st Bn. Doingt wood.
,, ,, H.Q. Templeux La Fosse	4th Bn. Buire wood.
,, ,, ...	5th Bn. Buire wood.
Third Army, 2nd Tank Brigade	2nd Bn. Velu wood.
,, ,, H.Q. Thilloy	8th Bn. Fremicourt.
,, ,, . . .	10th Bn. Fremicourt.
,, ,, 3rd Tank Brigade	6th Bn. Wailly.
G.H.Q. Reserve, H.Q. Henincourt	3rd Bn. Bray.
,, ,, . .	9th Bn. Bray.
,, ,, 5th Tank Brigade	13th Bn. Bray (unequipped).
,, ,, H.Q. Bray.	
First Army, 1st Tank Brigade	7th Bn. Boyeffles.
,, ,, H.Q. Bois d'Olhain	11th Bn. Bois des Alleux.
,, ,, ...	12th Bn. Bois de Verdrel.

It will be seen that by this date the Tank Corps had grown from three to five brigades, in all thirteen battalions; machines, however, were short, and the total fighting strength in tanks at this time was only 320 Mark IV and 50 Medium A Tanks (Whippets) fit for action.

The general plan was that tank units should co-operate with Army and Corps reserves in the deliberate counter-attack against tactical points in what was known as the battle zone, a belt of ground running several miles in rear of and parallel to the forward or outpost zone; no retirement from this zone was to be contemplated. Prior to March 20 the weather had been fine, the ground was good and a thorough reconnaissance had been made of some 1,500 square miles of country; supply dumps had been formed and communication by wireless, cable, dispatch rider and runner established throughout the units of the Tank Corps.

On March 21 at 5 a.m. the German bombardment opened on a front roughly running from La Fère to the River Scarpe, with a break round the old Cambrai battlefield. The first tanks to be engaged were three forward sections of the 4th Battalion north-west of St. Emilie, north-west of Peizière, and at Geninwell copse. These came into action about noon and fought most gallantly against heavy odds. The first section, supported by two companies of the 2nd Royal Munster Fusiliers, recaptured a battery of guns near Esclairvillers wood; later on in the day this section assisted in the counter-attack of the Connaught Rangers on Ronssoy wood; meanwhile the second section cleared the bridge and cutting north-east of Peizière.

Whilst these actions were being fought in the Fifth Army area, on the Third Army front one company of the 8th Tank Battalion co-operated with the 57th Infantry Brigade in a counter-attack on the village of Doignies. Zero hour was fixed at 6.40 p.m., but the attack was delayed and it was almost dark before the objective was reached. The village was cleared of the enemy, but on account of the darkness it was never completely occupied by our own men and eventually passed back into German hands.

On the following day, March 22, an Advanced Tank Corps Headquarters was opened at Hamencourt, a mile east of Doullens, in order to facilitate the battle liaison duties of the staff. On this day a most successful and gallant action was fought by the 2nd Tank Battalion in the neighbourhood of Vaux Vraucourt and Morchies. At 2.45 p.m. orders were issued for the 2nd Tank Battalion to advance and counter-attack the enemy, who had broken through the line Vaux Vraucourt-Morchies and was pushing forward towards Beugny. Two companies of infantry were detailed to support the tanks, but as eventually these could not be spared the tanks went into action alone.

The counter-attack began to develop around Beugny at about 4.30 p.m. Concentrated artillery fire was brought to bear on the tanks, but in spite of this they advanced amongst the enemy, put a field battery out of action, and by enfilading several trenches full of Germans inflicted heavy casualties on them. The enemy was eventually driven back behind the Vaux Vraucourt-Morchies position, which was then reoccupied by our infantry. Thirty tanks took part in this action; seventeen of these were hit and 70 *per cent.* of casualties suffered by their crews. Heavy though these losses were the enemy had suffered severely, and more important still his plan of action was upset.

On the Fifth Army front the penetration effected by the enemy caused a rapid withdrawal of our troops, and to cover this the 4th and 5th Tank Battalions moved eastwards on either side of the Cologne River, which joins the Somme at Peronne; the village of Epehy was cleared of the enemy and much valuable time was gained at Roisel and Hervilly by tank counter-attacks. The German infantry would not face the tanks, and broke whenever they saw them advancing.

On March 23 no tank action was fought on the Third Army front. On the Fifth Army front, the 1st Tank Battalion, which had not yet been engaged, took up a position on the reverse slope west of Moislains with machine-gun posts pushed out on the forward slope. The enemy, however, would not attack the line of tanks but worked round their flanks—the 1st Tank Battalion eventually withdrew towards Maricourt. The 4th and 5th Tank Battalions covered the withdrawal of our infantry on either side of the Cologne River, and by the evening ten tanks of the 4th Battalion had concentrated at Cléry and those of the 5th Battalion at Brie bridge, three miles south of Peronne. Shortly after their arrival here this bridge was blown up and the whole of the 5th Battalion tanks, except three, had to be destroyed for lack of petrol. Of these three, one succeeded in crossing the bridge after the explosion and the remaining two effected their escape *via* Peronne. All three were lost on the next day.

The following day, the 8th Tank Battalion was engaged in a most successful action south-east of Bapaume. Two companies advanced against Bus and Barastre, while a third covered the 6th Infantry Brigade's consolidation of a line of trenches. All tanks came into action and inflicted heavy casualties at close range, the enemy was checked for a considerable time and the 2nd Division was thus enabled to extricate itself from a most difficult position with little loss. The enemy was in force, but as was always the case, he would not face the tanks, and if he could not work round their flanks his advance halted until his guns could be brought up to deal with them.

It was on March 24 that a considerable number of Lewis-gun sections were first formed during this battle out of tank crews who had lost their machines. The 9th Tank Battalion handed its machines over to the 3rd Battalion and moved out as a Lewis-gun Battalion from Bray to assist the 35th Division and the 9th Cavalry Brigade in the defence of Montauban and Maricourt. The instructional staff of the Tank Driving School which, in February, had moved from Wailly to Aveluy, was rapidly formed into Lewis-gun sections, and with such tanks as were fit for action held a defensive line from Fricourt to Bazentin, covering the Albert-Bapaume road. The 5th Tank Battalion, south of the Somme, now without machines, was also formed into a Lewis-gun Battalion as crews were collected. This battalion in particular carried out most gallant and useful work, forty-five Lewis-gun groups being kept continuously in action. Several of these groups lost touch with their headquarters, but continued fighting with any troops in their vicinity until March 31.

On March 25 two companies of the 10th Tank Battalion came into action at Achiet-le-Grand and Achiet-le-Petit. At the first-named village, with the 42nd Division, one of these companies attacked the enemy, who, in large numbers, had broken through near Bapaume, and delayed his advance for several hours. By this date no fewer than 113 Lewis-gun groups had been posted in the La Maisonette—Chaulnes, Bray and Pozières—Contalmaison—Montauban—Maricourt areas, and during the night twenty more were sent out to hold the crossings over the River Ancre between Aveluy and Beaucourt. At this time Grandecourt and Miraumont were already in the enemy's hands and the position was most precarious. These groups held these crossings for several days and inflicted heavy casualties on the enemy each time he attempted to force a passage.

March 26 is an interesting date in the history of the Tank Corps, for, on the afternoon of this day, the Whippet Tanks made their debut. Twelve of these machines, belonging to the 3rd Tank Battalion, moving northwards from Bray were ordered to advance through the village of Colincamps to clear up the situation, which was very obscure. About 300 of the enemy were met with advancing on the village in several groups; these were taken completely by surprise, and on seeing the rapidly moving tanks fled in disorder, making no attempt at resistance. The Whippets then patrolled towards Serre and after dispersing several strong enemy patrols withdrew, having suffered no losses in tanks or personnel. This action was particularly opportune, as it checked an enveloping movement directed against Hebuterne at a time when there was a gap in our line.

Plate 3: Medium Mark "A" Tank (Whippet).

Save for a few minor tank engagements on March 27, 28, 29, 30, and 31, so far as the Tank Corps was concerned, the Second Battle of the Somme had come to an end, and, before closing this chapter, it is of interest to deduce the main lessons learnt from these the first defensive operations the Tank Corps had ever taken part in.

On March 21, tanks were too scattered ever to pull their full weight. To hit with them as they then were distributed was like hitting out with an open hand in place of a clenched fist, and when the blow fell there was no time to hit and simultaneously close the fingers. Out of a total of some 370 tanks only 180 came into action. The continual withdrawal of tanks by infantry formations in place of moving them forwards amongst the enemy resulted in many machines being worn out before they had fired a shot; this was a faulty use of an offensive weapon.

The two main lessons learnt were: firstly, that speed and circuit are the two essentials for an open-warfare machine; and secondly, one which has already been mentioned but which is so important that it is worth mentioning again, namely, that no great Army, such as the Germans massed against us on March 21, can depend on road and rail supply only. Consequently, unless these means of supply are supplemented by cross-country mechanical transport, that is, transport which is independent of road and rail, the greatest success will always be limited by the endurance of the horses' legs.

Men without supplies are an incumbrance, and guns and machine guns without ammunition are mere scrap iron. Had the Germans after March 26 been able to supply their troops mechanically across country, there can be little doubt that their advance would have been continued, for we could not have stopped it, and they might well have won the war. Fatigue may stop an advance gradually, but lack of supplies will stop it absolutely—this is the second and greatest lesson of the Second Battle of the Somme, if not of the entire war.

Chapter 24
Tank Signalling Organisation

In battle, co-operation between the commander and his troops, and between the troops themselves, depends very largely on the efficiency of the signal organisation. In a formation such as the Tank Corps, the chief duty of which was close co-operation with the infantry, the necessity for a simple though efficient communication was fully realised by Colonel Swinton as far back as February 1916, when he wrote his tactical instructions for the use of tanks, extracts from which have been given in Chapter 4. Though time for instruction was limited, special wireless apparatus was prepared and men trained in its use, but as orders were received not to equip the tanks with this apparatus, they were dispatched to France in August 1916 without it.

On September 11 the first instructions relative to tank signals were published with the Fourth Army operation orders; they read as follows:

From tanks to infantry and aircraft:
Flag Signals
Red flag	Out of action
Green flag	Am on objective
Other flags	Are inter-tank signals

Lamp Signals
Series of T's	Out of Action.
Series of H's	Am on objective.

A proportion of the tanks will carry pigeons.

The use made of these signals is not recorded, and no time was available, until after operations were concluded in November, wherein to organise more efficient methods.

In January 1917 steps were taken to introduce into the Heavy Branch some system of signalling in spite of the many difficulties, the chief of which were:

(1) No personnel other than the tank crews could be obtained.

(2) At most only two months were available for training.

(3) Neither the Morse nor semaphore codes could be read by infantry.

The whole question, after careful consideration, was fully dealt with in "Training Note No. 16," already mentioned.

The entire system of field signalling was divided under three main headings:

(1) *Local.*—Between tanks and tanks and tanks and the attacking infantry; also between the Section commander and the transmitting station, should one be employed.

(2) *Distant.*—Between tanks and Company Headquarters, selected infantry and artillery observation posts, balloons, and possibly aeroplanes.

(3) *Telephonic.*—Between the various tank headquarters and those

of the units with which they were co-operating.

The means of signalling adopted were as follows:

For local signalling coloured discs—red, green, and white. One to three of these signals in varying combinations could be hoisted on a steel pole. In all thirty-nine code signals could thus be sent, *e.g.* white = "Forward"; red and white = "Enemy in small numbers"; red, white, green = "Enemy is retiring." These codes were printed on cards and distributed to tank crews and to the infantry. Besides these "shutter signals" were also issued, but as they entailed both the sender and reader understanding the Morse code they were seldom used. The chief local system of communication was by runner, and it remained so until the end of the war.

Distant signalling was carried out by means of the Aldis daylight lamp, and as message-sending was too complicated a letter code was used, thus—a series of D.D.D. . . . D's meant "Broken down," Q.Q.Q. . . . Q's "Require supplies." Generally speaking, until November 1917, distant signalling was carried out by pigeons, which, on the whole, proved most reliable as long as the birds were released before sunset; at a later hour than this they were apt to break their journey home by roosting on the way.

In February 1917 Captain J. D. N. Molesworth, M.C., was attached to the Heavy Branch to supervise the training in signalling. This officer remained with the Tank Corps until the end of the war, and in 1918 was promoted to the rank of Lieutenant-Colonel and appointed Assistant Director of Army Signals in 1918. Under his direction classes in signalling were at once started and considerable progress was made in the short time available before the Battle of Arras was fought.

In this battle the various means of communication laid down were put to the test of practical experience. The telephone system was described by the 1st Tank Brigade Commander as "heart-breaking." "Many times, it was totally impossible to hear or to be heard when speaking to Corps Headquarters at a distance of five to six miles." Pigeons were most useful, the Aldis lamp was found difficult, and many messages were sent from tank to tank, and in some cases to infantry with good results, by means of the coloured discs.

The experiences gained pointed to the absolute necessity of allotting sufficient personnel to battalions for purposes of signalling and telephonic communication.

The result of these experiences was that in May the first Tank Signal Company was formed, the personnel being provided from those already trained in the tank battalions, to which a few trained Signal Service men were added. The formation of this company was shortly followed by that of the 2nd and 3rd Companies, the 2nd Company taking part in the Battle of Messines.

In May the first experiments in using wireless signalling from and to tanks were carried out at the Central Workshops at Erin, various types of aerials being tested. In July a wireless-signal officer was appointed to the Tank Corps and he at once set to work to get ready six tanks fitted with wireless apparatus for the impending Ypres operations.

These signal tanks, when completed, were allotted to the Brigade Signal Companies, and in isolated cases, during the battle, came into operation, but in the main they did not prove a great success on account of the extreme difficulty of the ground. Eventually these tanks were placed at different points along the battle front and were used as observation posts by the Royal Flying Corps, wireless being employed to inform the anti-aircraft batteries in rear whenever enemy's aeroplanes were seen approaching our lines. Many wireless messages were sent and much experience was gained by means of this work.

By the end of September, on account of signalling equipment being obtained, it was possible to carry out training on much better lines than heretofore. This was fortunate, for it enabled intensive signalling training to be carried out prior to the Battle of Cambrai. During this battle a much more complete system of signals was attempted, and wireless signalling proved invaluable in keeping in touch with rear headquarters and also in sending orders forward. On the first day of this battle a successful experiment in laying telegraph cable from a tank was carried out, five tons of cable being towed forward by means of sledges, the tank carrying 120 poles, exchanges, telephones, and sundry apparatus from our front line to the town of Marcoing.

The signalling experiences gained during the Battle of Cambrai proved of great value, the most important being that it became apparent that it was next to useless to attempt to collect information from the front of the battle line. Even if this information could be collected, and it was most difficult to do so, it was so local and ephemeral in importance as to confuse rather than to illuminate those who received it. Collecting points about 600 yards behind the fighting tanks were found to be generally the most suitable places for establishing wireless and visual signalling stations.

At these stations, officers were posted to receive messages and to compile them into general reports, which from time to time were transmitted by wireless to the headquarters concerned.

After the Battle of Cambrai, the 4th Brigade Signal Company was formed. This Company was the first one to have a complete complement of trained Signal Service officers and men allotted to it. It carried out exceptionally good work during the operations in March 1918.

At this time the complete organisation of signals in the Tank Corps may be shown graphically as follows:

A. D. SIGNALS
(Technical Instructions, Posting of Officers and Men, Control of all Signal Stores)

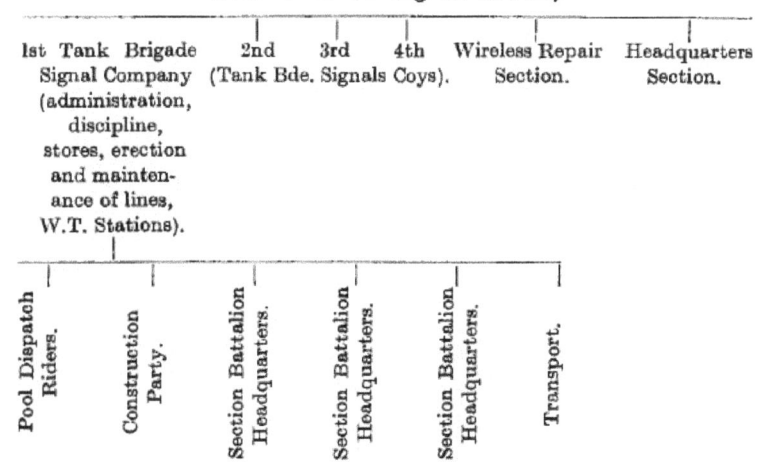

1st Tank Brigade Signal Company (administration, discipline, stores, erection and maintenance of lines, W.T. Stations).

2nd 3rd 4th (Tank Bde. Signals Coys).

Wireless Repair Section.

Headquarters Section.

Pool Dispatch Riders.

Construction Party.

Section Battalion Headquarters.

Section Battalion Headquarters.

Section Battalion Headquarters.

Transport.

Early in 1918 the type of wireless apparatus as used in the signal tanks was changed to C.W. (continuous wave) sets, these being more compact, and greater range of action being possible with the small aerials the tanks had to use.

Eight of these C.W. sets were issued to each Brigade Signal Company, and training in their use was carried out up to the commencement of the August operations. On the whole they proved a success and justified their adoption, but as experience was gained it became evident that something better and stronger was wanted.

In September a scheme was devised whereby the entire signal organisation of the Tank Corps was to be recast so as to fit in with the new tank group system, which was then being worked out for 1919. This organisation included Group Signal Companies and much larger Brigade Signal Companies than had hitherto been used, and the main type of apparatus that this organisation was to use was wireless. Only one set of wireless to each tank company was to be employed actually in tanks, the other stations being carried forward in box cars so as to render them more mobile.

The importance of signalling in a formation such as the Tank Corps cannot be over-estimated, and this importance will increase as more rapid-moving machines are introduced, for, unless messages can be transmitted backwards and forwards without delay, many favourable opportunities for action, especially the action of reserves, will be lost. Making the most of time is the basis of all success, and this cannot be accomplished unless the commander is in the closest touch with his fighting and administrative troops and departments.

CHAPTER 25
The French Tank Corps

The existence of the French Tank Corps was due to the untiring energy of one man—Colonel (now General) Estienne. On December 1, 1915, this officer, then commanding the 6th French Divisional Artillery, addressed a letter to the Commander-in-Chief of the French Armies in which he expressed his firm belief that an engine of war, mechanically propelled and protected by armour, capable of transporting infantry and guns, was the solution to the deadlock on the Western Front. The idea of the machine in Colonel Estienne's mind was the result of his work throughout the year 1915, during which period he had seen Holt tractors in use with British artillery units.

On December 12, 1915, Colonel Estienne was given an interview at G.Q.G., the French General Headquarters, where he set forth his theory of mechanical warfare. On the 20th of this month he visited Paris to discuss the details of his machine with the engineers of the Schneider firm; but it was not until February 25, 1916, that the Department for Artillery and Munitions decided to place with this firm an order for 400 of these armoured vehicles.

Meanwhile, Colonel Estienne returned to his command, the 3rd Corps Artillery, before Verdun, but still kept in unofficial touch with the manufacturers. Two months later he learned that a similar number of cars, but of a different pattern, were to be made by the St. Chamond works. These machines were of a heavier type with a petrol-electric drive.

In June 1916 the French Ministry of Munitions, which had meanwhile been created, decided on an experimental and instructional area at Marly-le-Roi. Later on, a depot for the reception of stores was established at Cercottes. On September 30, Colonel Estienne was promoted to the rank of General and gazetted "*Commandant de l'Artillerie d'Assaut aux Armées*" and was appointed the Commander-in-Chief's delegate to the Ministry of Munitions in matters connected with tanks; he thus became the official connecting link between the armies in the field and the constructional organisation of the Ministry.

In October a training centre was established at Champlieu on the southern edge of the forest of Compiègne, and it was here that the first tank units were assembled on December 1, 1916. During the succeeding months, Schneider (see Plate 4) and St. Chamond (see Plate 4) machines continued to arrive, and training was carried out at this camp until the German offensive of 1918.

On June 20 a tank establishment was sent to the Ministry of Munitions and was approved of a month later. This establishment comprised four Schneider battalions and four St. Chamond battalions, and the creation of two tank training centres besides Champlieu, namely, Martigny and Mailly Poivres.

Meanwhile, General Estienne in June visited England, and having seen the British Mark I machine was convinced of the necessity of a lighter tank. This tank was the result of an idea he had in mind, namely, of producing on the battlefield waves of skirmishers in open order; each skirmisher to be clad in armour, and to be armed with a machine gun which could be used with uninterrupted vision in all directions. The weight of armour necessitated an auxiliary means of motion; this, in its turn, gave rise to the necessity for another man to drive the machine. These views General Estienne laid before the Renault firm in July 1916, and at the same time he urged the Ministry to accept his proposed light tank, but without success.

Complete designs were, however, prepared and on November 27 General Estienne was able to propose to Marshal Joffre the construction of a large number of light tanks for future operations and to inform him of the existence of the design of such a machine; in fact, 150 had already been ordered as "Command" tanks for the heavy battalions (see Plate6). Still the Ministry was not convinced, and it was not until further trials had taken place that, in May 1917, an order for 1,150 was authorised. This number was increased in June to 3,500, when a new sub-department of the Ministry of Munitions known as "*Le Sous-Direction d'Artillerie*" was formed to deal with the production and design of tanks.

In spite of all General Estienne's endeavours, he was still experiencing from certain adherents of the old school, the thinkers in "bayonets and sabres," that unbending opposition which had proved so formidable an antagonist to the progress and expansion of the British Tank Corps, and it was not until the Battle of Cambrai had been fought, in November 1917, that the French Ministry of Munitions was finally convinced of the value of the tank. Opposition now ceased, and in order to accelerate the output, the firms of Renault, Schneider, and Berliet were all engaged in the manufacture of light *chars d'assauts*.

In December 1917 it was decided to form 30 light tank battalions of 72 fighting and 3 wireless signal machines each. Of these 30 battalions 27 were in the field and the remaining 3 undergoing their preliminary training at Cercottes on the date of the signing of the armistice.

The operations of the French Tank Corps may be divided into three well-defined periods:

(1) First period, 1917, birth and infancy of the Schneider and St. Chamond types.

(2) Second period, first half of 1918, adolescence and maturity of the Schneider and St. Chamond, and the infancy of the Renault type.

(3) Third period, second half of 1918, adolescence and maturity of the Renault machine.

During the first period three battles were fought:

On April 16, 1917, the French tanks fought their first engagement, taking part in the operations of the Fifth French Army in the attempted penetration on the Chemin des Dames. Eight Schneider companies were employed. Three of these were to operate between the Craonne Plateau and the Miette, and five between the Miette and the Aisne. The former companies failed to get into action and suffered heavy losses from the enemy's artillery, which from the heights of the Craonne plateau commanded their advance.

The latter companies succeeded in crossing the second and third lines of the enemy's defences, but in spite of their remaining for a considerable time in front of the infantry these troops could not follow owing to the enemy's heavy machine-gun fire. At nightfall the tank companies were rallied, having sustained serious losses in personnel and *matériel*. Bodies of infantry had been specially detailed to escort the tanks and prepare paths for their advance, but their training had been limited and their efforts were ineffectual.

On May 5 one St. Chamond and two Schneider companies took part in a hurriedly prepared operation with the Sixth Army. The

Plate 4

French Schneider Tank

French St. Chamond Tank

Schneider companies led the infantry in a successful attack on Laffaux Hill, and of the sixteen St. Chamond tanks detailed for the action only one crossed the German trenches.

Between May and October preparations were made by the Sixth French Army for an attack on the west of the Chemin des Dames, and for this attack infantry were trained with the tanks at Champlieu and special detachments, known as *troupes d'accompagnement*, were instructed in the ways and means of assisting the tanks over the trenches.

The attack, which became known as the Battle of Malmaison, was fought on October 23. Five companies of tanks took part in it under the orders of Colonel Wahl, who had recently been appointed to command the *Artillerie d'Assaut* with the Sixth Army. This command was the origin of what later became a Tank Brigade Headquarters, which corresponded with a Group Headquarters in the final organisation of the British Tank Corps.

In this battle the Schneider company operated with success, but the St. Chamond machines were a failure, only one or two reaching the plateau. On the 25th the St. Chamonds were used again.

Generally speaking, it was considered that the French heavy tanks had justified their construction, nevertheless many still doubted their utility when the victory of Cambrai, on the British front, dispelled all doubts in the French mind.

The second period now opened and defensive reconnaissances were undertaken along the French front in view of the expected German offensive.

In March 1918 all available tanks were concentrated behind the front of the Third French Army as counter-attack troops, and in this capacity took part in the following minor operations, which were chiefly undertaken to recapture features of local tactical importance: on April 5 at Sauvillers; on April 7 at Grivesnes; on April 8 at Sénécat wood, and on May 28 at Cantigny in co-operation with American troops.

Following the great blow struck at the junction of the British and French Armies in March the German General Staff decided to attack the French on May 27. It would appear that this attack was at first intended only to secure the heights south of the River Vesle, but that by the 29th, owing to its astonishing initial success, it was decided to push it forward with the ultimate intention of capturing Paris and so ending the war before America could develop her full strength. In support of this intention there is evidence that a council of war was held in the recaptured area at which the *Kaiser*, Crown Prince, Hindenburg and Ludendorff were present and at which it was decided to exploit the success gained to its utmost, not, however, losing sight of the original plan, which was to include the capture of Reims. This offensive may be considered to have worn itself out by June 4, on which date the Germans had developed a salient forty kilometres deep on a forty kilometres front. The old capital of France, however, remained in French hands and its occupation denied to the German forces holding the salient a most needed line of supply.

On June 9 the attack was extended, being directed against the Third French Army between Noyon and Montdidier. Behind this Army four heavy tank battalions had been assembled. The first and second lines soon fell into the enemy's hands, and the French troops, which had been detailed for counter-attack, were rapidly absorbed in the defence. On the 10th reinforcements were hurried forward, and on the 11th General Mangin launched his tank and infantry counter-attack. This battle continued until the 13th, and in spite of the many difficulties 111 out of the 144 tanks assembled started at zero hour. Losses in machines were heavy and about 50 *per cent.* of their crews became casualties, but in spite of this and the fact that the tanks rapidly outdistanced the infantry, a heavy blow was inflicted on the enemy, whose offensive definitely broke down.

In the action of June 11, the Schneider and St. Chamond tanks reached the zenith of their career. From now onwards, though they continued to be fought, they gradually ceased to be used as units, becoming mixed with Renault machines until finally, in October 1918, the two remaining mixed battalions were armed with British Mark V star tanks; these two battalions, however, never took the field.

In order to stop the enemy's onrush on May 27, two battalions of Renault tanks were hurried up by road to the north-eastern fringes of the forest of Villers-Cotterets, and on May 31 they made their debut, two companies co-operating with colonial infantry on the plateau east of Cravançon farm. From this date on to June 15, these two battalions continued to act on the defensive with tired troops; nevertheless, they succeeded in preventing a further advance of the German Armies. This closes the second period.

During the first fortnight of July the 3rd and 5th Renault Battalions were moved to the battle area, the former being attached to the Fifth French Army, south of Dormans, and the latter to the Tenth. These machines came into action on the 15th, 16th, and 17th of the month.

On July 15 the Germans launched their final great attack of the war, the blow falling between Château-Thierry and Reims. The French Armies involved in this battle were holding the following sectors:

(1) The Tenth Army, between the Aisne and the Ourcq.
(2) The Sixth Army, between the Ourcq and the Marne.
(3) The Fifth Army, between the Marne and Reims.
(4) The Fourth Army, east of Reims.

The warning order to concentrate his units was received by the G.O.C., French Tank Corps, on July 14. At that time the G.O.C. Tenth French Army had at his disposal five heavy battalions and three light, and the Fifth and Sixth French Army respectively now received one heavy and three light battalions. The total number of tank battalions available was, therefore, seven heavy battalions and nine light ones.

The main attack was to be made by the Tenth French Army, whilst the Sixth and Fifth Armies were to intervene, when the time was ripe, in order to harry the enemy in a retirement which would be inevitable if the attack of the Tenth Army was successful. The entire operation was to be based on tanks, which were to be engaged to the last machine. As this was the greatest French tank battle fought during the war it is interesting to enter, in some detail, into the operations of the tanks allotted to the French Tenth Army.

On July 14, when orders were issued for the concentration of tanks on the Tenth French Army front, Colonel Chedeville, commanding the 2nd Tank Brigade, was with the Third French Army. He had at his disposal three St. Chamond battalions, the 10th, 11th, and 12th, two Schneider battalions, the 3rd and 4th, and one complete Light Brigade comprising the 1st, 2nd, and 3rd Renault battalions, and the 1st Schneider battalion. Of these the first five were in the First and Third French Army areas, and had suffered severely in the counter-attack of June 11. Having received his orders, Colonel Chedeville at once assembled his battalion commanders and explained to them the situation. At 6 p.m. a further conference was held at which the proposed sectors of attack were allotted for reconnaissance. These reconnaissances were completed by 6 p.m. on the following day, and on them was based "Army Operation Order No. 243," in which tank units were allotted as follows:

1st Corps

3rd Heavy Tank Battalion . Allotted to 153rd Division.
(27 tanks).

XXth Corps

12th Heavy Tank Battalion . Allotted to 2nd American Division.
(30 tanks)
11th Heavy Tank Battalion   ,,   ,,   2nd American Division.
(30 tanks)
4th Heavy Tank Battalion   ,,   ,,   Moroccan Division.
(48 tanks)
1st Heavy Tank Battalion .   ,,   ,,   1st American Division.
(48 tanks)

XXXth Corps

10th Heavy Tank Battalion . Allotted to 38th Division.
(24 tanks)

In Army Reserve in the region of Villers-Cotterets—Fleury

1st Light Tank Battalion (45 tanks)
3rd Light Tank Battalion (45 tanks)
2nd Light Tank Battalion (40 tanks)

The assembly positions from north to south of the various units were as follows:

3rd Heavy Tank Battalion	Ravine south-west of Montigny-Lengrain.
12th Heavy Tank Battalion	Ravine north of Mortefontaine.
11th Heavy Tank Battalion	Two Companies Ravine Longavesne and Lepine farm, 1 Company ravine 1 kilometre north of Soucy.
4th Heavy Tank Battalion	Northern fringes of the forest of Villers-Cotterets south-east of Vivières.
1st Heavy Tank Battalion	Maison Forestière, 200 metres north of the railway on the road from Villers to Soissons.
10th Heavy Tank Battalion	Cross-roads south-east of the Cordeliers cross in Villers-Cotterets forest.
1st Light Tank Battalion	Northern edge of Villers-Cotterets forest, south-west of Vivières, ready to attack in the wake of the Moroccan Division.
2nd Light Tank Battalion	Northern edge of forest south-west of Vivières ready to follow 2nd American Division.
3rd Light Tank Battalion	St. George's cross, ready to support either the 48th Division or the XIth Corps.

Owing to the failure of the Military Transportation Authorities great delay was occasioned in the arrival of several units, and in some cases, tanks had to be left behind. Generally speaking, detraining stations were not far enough forward; this resulted in the 1st and 3rd Light Battalions arriving late at their destinations.

During the night of July 17–18, the various units proceeded to their starting-points in rear of their respective lines of attack.

3rd Heavy Tank Battalion 153rd Division	} St. Bandry—Saconin et Breuil—Vauxbuin.
11th Heavy Tank Battalion 12th Heavy Tank Battalion 1st American Division	} Cutry—Missy aux Bois Ploisy.
4th Heavy Tank Battalion Moroccan Division	} St. Pierre Aigle—Chaudun—Villemontoire.
1st Heavy Tank Battalion 2nd American Division	} Chavigny—Beaurepaire Forest—Vierzy—Tigny.
10th Heavy Tank Battalion 38th and 48th Divisions	} Longpont—Villers Helon—Le Plessier Huleu.

The attack was launched at 4.35 a.m. in a slight fog which accentuated its surprise. There was no artillery bombardment. At 7.30 a.m., owing to the difficulties in communication and the rapidity of the advance, the Light Tank Battalions in Army reserve were placed at the disposal of the XXth and XXXth Corps in order to support the Divisions which had penetrated the deepest.

In this attack the enemy's resistance was not unusually stubborn and the tanks and infantry advanced to a considerable depth without difficulty. Several tanks of the 12th Heavy Battalion fell out by the way, but those of the 10th succeeded beyond expectation in negotiating the difficult ground in the neighbourhood of Longpont. Of the Renault battalions only the first came into action, being launched at 7 p.m. in an attack on Vauxcastille ravine in which it succeeded in leading the infantry forward to a depth of three to four kilometres.

Of the 324 tanks which were concentrated in the Tenth French Army Sector, 225 were engaged on July 18. Of these 102 became casualties, 62 being put out of action by artillery fire. In personnel the losses were about 25 *per cent.* of the effectives engaged.

On July 19, composite units were formed and 105 machines took part in this day's fighting, which consisted in divisional attacks on limited objectives launched at various hours during the day. By now the enemy's resistance had increased so much that several of the tank battalions suffered heavily. The 3rd Heavy Battalion had, by the end of the day, lost all its remaining tanks save two, but in sustaining these casualties it had pushed the line forward to the Chaussée Brunehaut. In the 12th Heavy Battalion only one machine reached its final objective. In spite of this severe resistance the attack was a great success. Of the 105 tanks operating fifty were hit by shell fire, and casualties

BATTLE OF SOISSONS
July 18th 1918.

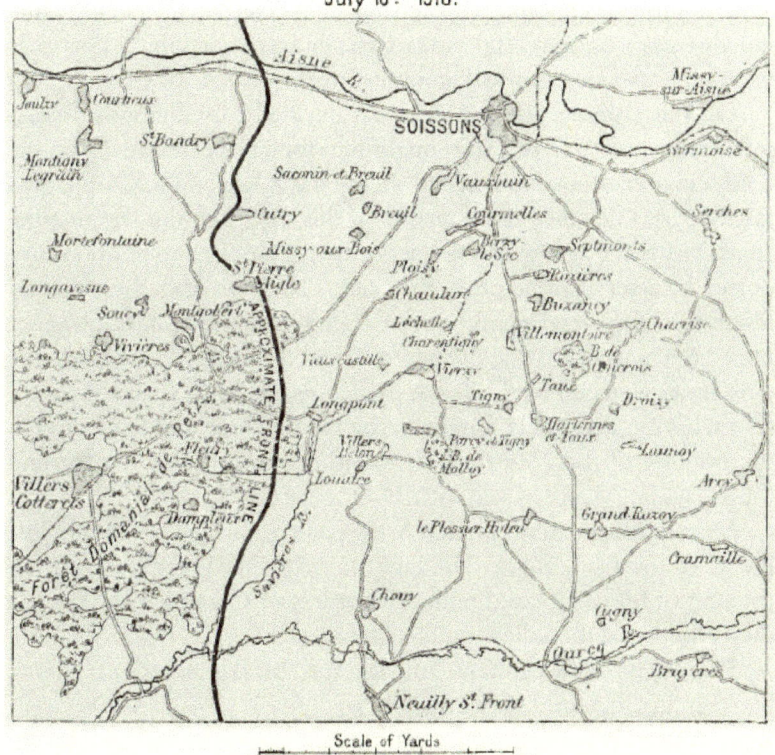

amongst crews totalled up to 22 *per cent.* of the personnel engaged.

On the following day only small local counter-attacks were carried out; in these thirty-two tanks took part, of which seventeen were hit and no less than 52 *per cent.* of their crews became casualties.

On July 21 the XXth Corps carried out a prepared attack, the first objective being the line Buzancy—eastern edge of Concrois wood—Hartennes wood, and the second the line of Chacrise. The attack was launched without artillery preparation and the villages of Tigny and Villemontoire were captured, but later on retaken by the enemy. During this day's fighting 100 tanks were engaged, of which thirty-six were hit; losses in personnel amounted to 27 *per cent.* of effectives.

On the evening of the 21st it was decided to withdraw all tanks into Army reserve so that they might refit for a projected attack on the 23rd. This attack was launched at 5 a.m., the XXth and XXXth Corps taking part. The chief characteristic of this day's fighting was that the attack was made against an enemy occupying a defensive position supported by a very strong force of artillery. The result of this was that no fewer than forty-eight tanks out of eighty-two were hit. It, however, must be remembered that during the six succeeding days of battle the tank units, attached to the Tenth French Army, had exhausted themselves, having practically fought to the last machine and last man. On the evening of the 23rd they were withdrawn in Army reserve, and three days later were placed in G.H.Q., reserve.

Meanwhile the Sixth French Army had conformed to the requirements of the main attack. The tank units of this Army were, on the evening of July 14, placed under the orders of Commandant Michel; they comprised the following battalions:

503rd Renault Regiment: 7th, 8th, and 9th Battalions.

13th St. Chamond Battalion.

On July 15, company commanders reconnoitred the front of attack, the tanks meanwhile being got ready for entrainment. On July 18 all units were in position with the infantry units to which they had been allotted, as follows:

7th Light Battalion	2 Companies to the 2nd Division.	
8th ,, ,,	3 ,, ,,	47th ,,
9th ,, ,,	1 ,, ,,	164th ,,
	2 ,, ,,	63rd ,,
13th Heavy Battalion	1 ,, ,,	47th ,,
	2 Companies in Army reserve.	

The 2nd and 47th Divisions were in the IInd Corps, whilst the 63rd and 164th Divisions were in the VIIth Corps.

At zero hour plus thirty minutes the tanks left their starting-points. The 7th and 8th Light Battalions operated effectively in the capture of the heights west of Neuilly St. Front and hill 167. The attack of No. 325 Company of the 9th Light Battalion, operating with the 47th Division, was brilliantly executed north of Courchamps.

In the evening the tanks rallied, the attack being continued with all available machines on the following morning.

As a general rule a section of five tanks was affiliated to each attacking battalion. This policy continued to the end of the operations on July 26, when the regiment was withdrawn to rest, worn out more by "trekking" than by fighting. The casualties in this sector were extremely light.

When the front of the attack, launched by the Germans, on July 15, became known to the French Higher Command, a Light Regiment of tanks, consisting of the 4th and 6th Battalions, was hurriedly dispatched from the Sixth French Army area to the Fourth Army east of Reims. The 5th Battalion engaged one company with the 73rd Infantry Division of the Sixth French Army in the recapture of Janvier wood, south of Dormans, on July 15, and two companies on July 16 and 17, in "mopping up" in the direction of Bois de Conde, east of Château-Thierry.

When it was realised that the German attack east of Reims had failed, the 4th and 6th Battalions were hurriedly transported by road, between July 16 and 19, south of the Marne, south-west of Reims, to take part in local counter-attacks. These attacks were entrusted to the Ninth French Army, which had taken over command of all French troops south of the Marne, and had at its disposal the 4th, 5th, and 6th Light Tank Battalions, and two companies of heavy tanks, which had been rapidly sent up by train from St. Germaine between Epernay and Reims.

Two sections of the 4th Light Battalion were engaged on July 18 with two battalions of the 7th Infantry Regiment; two on July 20, with the 97th and 159th Regiments; and one on the 19th, with the 131st Division—all in the neighbourhood of the Bois de Leuvrigny south of the Marne. Later, on July 23, sections of the 4th Battalion were employed with British troops—the 186th Infantry Brigade in the attack on Marfaux and with the 56th and 60th Battalions of the *chasseurs-à-pied* at Connetreuil, whilst, on the same date, two sections of the 6th Battalion attacked with units of the 15th British Division between Espilly and Marfaux, and two more were employed unsuccessfully with the 37th Infantry Regiment against Fauants farm.

So ends the account of the tank actions in the Battle of Soissons.

This great victory, from a tank point of view, had a stupendous influence on succeeding operations, owing to:

(1) The eagerness with which Infantry Commanders now clamoured for tanks.

(2) The speeding up of the formation and training of new tank battalions.

From this date on, battalions of Renault tanks became available at the rate of one a week; this resulted in tired battalions being speedily replaced by fresh ones, consequently they were never so completely worn out as was the case in the British Tank Corps, which only received two fresh battalions between August and November 1918, one of which arrived too untrained ever to go into action.

The operations from now on will be very briefly described, as space does not permit of elaboration. It is, however, of interest that these tank actions should be enumerated, for they show that, without the assistance of the tank, a deadlock would have re-occurred.

On August 1, 45 French tanks took part in an engagement at Grand Rozoy. Then came the great British tank attack of August 8, in which the First and Tenth French Armies co-operated, 110 French tanks taking part on this day and the following, 80 advancing with the infantry a distance of 18 kilometres on the south of the Roye-Amiens road, whilst 30 made a 5-kilometre advance near Montdidier. Between August 16 and 18 the attack developed west of Roye; here 60 Renault and 32 Schneider machines were engaged; co-operation with the infantry was, however, difficult on account of the broken nature of the old battlefield across which the attack was now being pushed.

The next operation was a continuation of the Tenth Army's offensive; it took place between the Oise and the Aisne, beginning on August 20, and being continued intermittently up to and including September 3. On the 20th and 22nd, 12 Schneider, 28 St. Chamond, and 30 Renault tanks were engaged north of Soissons.

During the week commencing August 28, three Light Battalions advanced five kilometres between the Aisne and the Aillette, 305 machines being employed at different times during these operations.

The next operation in which tanks were engaged was the cutting off of the St. Mihiel salient, French tanks being used with the Second French and American Armies. During the two days' fighting, September 12 and 13, some 140 tanks took part in the battle.

On September 14, the Tenth French Army resumed its offensive east of Soissons, eighty-five Renault tanks co-operating between the 14th and the 16th. Ten days later an extensive joint attack was made by the Fifth and Second French Armies in conjunction with the American Army commencing on the 26th; this attack continued until October 9.

The Fourth Army attacked on a 15-kilometre front in the Champagne, and in all 630 Renault and 24 Schneider actions were fought. Meanwhile the Second French Army and the American Armies attacked on a 12-kilometre front between the Argonne and the Meuse, and advanced during the seven battle days some 15 kilometres; 350 Renault, 34 Schneider, and 27 St. Chamond actions were fought in connection with this advance.

At the urgent request of the Sixth French Army Commander, whose command had joined the "Grand Army of Flanders" after its work in the Soissons area had been concluded in July, a Renault battalion, less one company, and some heavy tank units were entrained for Dunkerque, the third company of this battalion having already been sent on detachment to Salonika at the urgent request of General Franchet d'Esperey. On September 30 and October 3 and 4, 55 tanks were employed north-west of Roulers, and from the 14th to the 19th, 178 tank engagements were fought, in which the enemy was driven back some 15 kilometres. This advance was continued on the 31st of this month in the direction of Thielt, and on this and the two following days 75 tank engagements took place.

From the end of September onwards, operations generally had consisted in following up the enemy all along the line and pressing back his rear-guards. On September 30, a minor tank action was fought between the Aisne and the Vesle; on October 16 another on the eastern bank of the Meuse, and between October 17 and 19 yet another north-east of St. Quentin, in co-operation with the British attack further north. In this last attack the French Army advanced ten kilometres on a three-kilometre front. The last actions fought by French tanks took place between October 25 and 31, the first south of the Oise and in the direction of Guise, when, on a front of five kilometres, an advance of no fewer than fifteen was made, the second north-west of Rethel, and the third north of Cruyshantem in Flanders.

In conclusion, it is interesting to summarise the statistics available and compare them with those of the British Tank Corps given at the end of Chapter 37.

In August the strength of the French Tank Corps was 14,649 all ranks, approximately the strength of an infantry division. During 1918, 3,988 individual tank engagements were fought: 3,140 by Renault, 473 by Schneider, and 375 by St. Chamond tanks. Tanks were employed on 45 of the 120 days which elapsed between July 15 and November 11. In personnel the casualties between these dates were approximately 300 officers and 2,300 other ranks.

Finally it may be stated that as there can be no doubt that July 18 was the second greatest turning-point in the war on the Western Front, the first being the Battle of the Marne in 1914, so can there be no doubt that the Battle of Soissons would never have been won had not the French possessed a powerful force of tanks whereby to initiate success. The German General Staff, which should be the best judge of this question, candidly admit that the French victory was due to the use of "masses of tanks." Neither was the General Commanding-in-Chief of the French Armies reticent, for on July 30 he issued the following special order of the day to the French Tank Corps: "*Vous avez bien mérité la patrie,*" whilst General Estienne, to whom so much was due, received the Cravat of the *Légion d'Honneur* and was promoted to the rank of General of Division for the great services he had rendered to his country.

Chapter 26
Preparations for the Great Offensive

As soon as the position resulting from the great German attack of March 21 began to stabilise steps were taken by the Headquarters of the Tank Corps to reorganise and refit its battalions. This work was most difficult on account of the reopening of the German offensive in the Lys area, which necessitated converting the 4th Tank Brigade into a Lewis-gun unit and dispatching it north to assist in stemming the German advance. Besides this, towards the middle of April instructions were received that on account of the difficulty of finding the required number of infantry reinforcements the number of tank brigades was to be reduced from six to four; this meant the disbanding of the 5th Brigade in France and the breaking up of the 15th, 16th, 17th, and 18th Tank Battalions in England. Yet another difficulty was the question of re-arming, many machines had been lost during the retirement, nevertheless, on account of insufficient transport, it was not found possible to ship out to France the new Mark V tanks, the production of which in England was now in full swing.

All these difficulties were eventually overcome, with the result that during June and July four brigades of the Tank Corps were re-armed; but before this question is dealt with it will be necessary to hark back to the various operations which bridge the period between April 4, the date upon which the Second Battle of the Somme ended, and August 8, when our own great offensive was begun.

On April 9 the Germans launched their second great attack between Festubert and Fleurbaix against the British front. It succeeded, so it is thought, even beyond their expectations, and by the 11th the enemy's line roughly ran as follows: East of Ploegsteert—Armentières—Steenwerck—Estaires—Lestren—Vieille Capelle, through Festubert to Givenchy, with the apex of the salient near Nieppe. This attack in all probability was meant as a feint directed against a weak spot in our line in order to threaten the coalfield round Bruay and so cause Marshal Foch to weaken his reserves in Champagne and on the Somme. Succeeding as it did at first, it appears that the German command attempted to develop it from a feint to a decisive attack, with the result that their own reserves, of which they had none too many, were involved as well as those of the Allies.

In order to meet the requirements of the new situation on April 11 detachments of the 7th and 11th Battalions of the 1st Tank Brigade were dispatched to hold a line west of Merville. Two days later the 4th Tank Brigade, now consisting of the 4th, 5th, and 13th Battalions, was turned into a Lewis-gun Brigade. On the 13th the 5th Battalion moved to Berthen, on the 16th the 13th Battalion to Boescheppe, and on the 17th the 4th Battalion to the same place.

By April 17 the distribution of the Tank Corps was as follows:

1st Tank Brigade	11th Battalion	N.E. of Busnes.
H.Q. Bois D'Ohlain	7th ,,	Molinghem.
	12th ,,	Simencourt.
2nd Tank Brigade	6th Battalion	Bailleulval.
H.Q. Saulty	10th ,,	La Cauchie.
3rd Tank Brigade	3rd Battalion	Toutencourt.
H.Q. Molliens-au-Bois	9th ,,	Merlimont.
	1st ,,	Frechencourt.
4th Tank Brigade	4th Battalion	Boescheppe.
H.Q. Godewaersvelde	5th ,,	Berthen.
	13th ,,	Boescheppe.
5th Tank Brigade	2nd Battalion	Blangy.
H.Q. Monchy Cayeux	8th ,,	Humières.

The fighting carried out by the Lewis-gun units was of a severe nature, so much so that the casualties sustained caused the greatest anxiety at the Tank Corps headquarters, as reinforcements from England were exceedingly limited; further, as it was still hoped to save the battalions at home to the Corps, it was especially desirable not to call upon them for drafts.

Early on April 24 the enemy attacked south of the River Somme on a front from Villers-Bretonneux to the Bois de Hangard. This attack is of special interest as it was the first occasion upon which the Germans employed tanks of their own manufacture against us.

******

The German reports published in April asserted that tanks were used against the British Army on March 21. As nothing is definitely known of their effect they probably failed to come into action.

******

By means of these tanks the enemy penetrated our front, captured most of the extensive village of Villers-Bretonneux and advanced as far as the Bois de l'Abbé. Prior to this attack, at 1 a.m., a section of tanks of the 1st Battalion, hidden in the Bois de l'Abbé, moved east of the wood owing to the excessive gas shelling. At 8.30 a.m. this section, under the orders of the 23rd Infantry Brigade, moved forward to secure the Cachy switch trench against the enemy's threatening attack; exactly an hour later two of our machines, both females, came into view of a hostile tank and were put out of action by its gunfire—it should be remembered here that female tanks are armed only with machine guns. Shortly afterwards a British male Mark IV machine hove into sight, and speeding into action there then took place the first tank *versus* tank duel to be recorded in history. This male soon scored a direct hit on its antagonist, whereupon the enemy evacuated their tank and fled. By this time three more enemy tanks had appeared; these the Mark IV machine engaged, and was in the process of driving off the field of battle when it received a direct hit from a field-gun shell and was put out of action.

South-west of Villers-Bretonneux, seven Whippet machines were sent out at 10.30 a.m. to clear up the situation east of the village of Cachy. Whilst proceeding round the north-east side of this village they suddenly came upon two battalions of Germans massing in a hollow preparatory to making an attack. Without a moment's hesitation the seven Whippets formed line and charged down the slope right on to the closely formed infantry. Indescribable confusion resulted as the Whippets tore through the German ranks, the enemy scattered in all directions, some threw themselves on their knees before the machines, shrieking for mercy, but only to be run over and crushed to death. In a few minutes no fewer than 400 Germans were killed and wounded. The Whippets, having now completed their task, *viz.* "clearing up the situation," returned, one machine being put out of action by artillery fire on the journey home; in all only five casualties amongst the crews were suffered during this action.

The two most remarkable features of this little engagement are: firstly, the helplessness of some 1,200 infantry against seven tanks manned by seven officers and fourteen other ranks; and, secondly, that the tanks left their starting-point, which was 3½ miles from the scene of action, at 10.30 a.m., covered ten miles of ground, fought a battle, and were back home again at 2.30 p.m.

On April 25 further minor tank operations took place in the Villers-Bretonneux area, chiefly east of the Bois d'Aquenne and the Monument, and, on the next day, British tanks for the first time in their history co-operated with the French Army, four tanks of the 1st Battalion being ordered to assist the Moroccan Division in an attack on the Bois de Hangard. This attack was not a success, due to two quite exceptional reasons; two trees were cut down during the night, which were to have acted as landmarks for the tanks, and the smoke barrage was in error put down to the east instead of the west of the German line; consequently the tanks not only lost their direction but were subjected to an intense machine-gun fire when nearing the German position; this prevented the French infantry co-operating with them.

The month of May was chiefly spent in re-sorting the tank battalions and resting the men. The embargo on the importation of tanks from England had now been removed, and Mark V tanks were arriving in France at the rate of sixty a week. This machine, very similar in shape to the Mark IV or Mark I, was a great improvement on all former types, it being a much more mobile and handy weapon. A new system of tactics was at once got out to cover its increase in power, and training was started so as to accustom all ranks to its use.

At about this time a considerable number of French troops were billeted in and around the Tank Corps area and it is a pleasure to record their extreme keenness to learn all they could about tanks and their tactics. General Maistre, commanding the Tenth French Army, with its headquarters then at Beauval, particularly asked that tank demonstrations should be held for the units of his command. This was done, and right through May and June two or three of these demonstrations were given weekly. Besides French troops, units from the Ist, XIth, XIIIth, XVIIth, and XVIIIth English Corps and the Canadian and Australian Corps also attended, the greatest benefit resulting to all taking part.

From the beginning of June onwards preparations were set on foot to have all tank units ready by August 1 for any eventuality. This necessitated intensive training, re-arming and re-equipping. Sledges for supply haulage were prepared, bridges for the passage of light tanks over wide trenches were made, cribs were constructed for the heavy tanks—these were large hexagonal crates which served the same purpose as the tank fascines did at the Battle of Cambrai; wire-pulling apparatus was got ready, smoke apparatus ordered, and portable railway ramps made. It was altogether an excessively busy time on the training ground and in the workshops, and, as matters eventually turned out, it was extremely fortunate that this work was taken up at this early date, for, as a future chapter will show, when the Tank Corps was next called upon to make ready for an extensive operation only eight days were obtainable to prepare in.

CHAPTER 27
The Battles of Hamel and Moreuil

During June and July three tank actions were fought: the first was a night raid on June 22–23, the second the Battle of Hamel, and the third the Battle of Moreuil or Sauvillers.

The night raid is interesting in that it was the first occasion in the history of the Tank Corps in France upon which tanks were definitely allotted to work at night. The raid was carried out against the enemy's defences near Bucquoy by five platoons of infantry and five female tanks. Its object was to capture or kill the garrisons of a series of posts. The raid took place at 11.25 p.m. A heavy barrage of trench-mortar and machine-gun fire was met with at a place called Dolls' House in "No Man's Land": here the infantry were held up, and though reinforced were unable to advance further. The tanks, thereupon, pushed on and carried out the attack in accordance with their orders.

It is worthy of note that not a single tank was damaged by the trench-mortar barrage, which was very heavy. The tanks encountered several parties of the enemy and undoubtedly caused a number of casualties. One tank was attacked by a party of the enemy who were shot down by revolver fire; later on, this tank rescued a wounded pla-

Plate 5: Mark V Tank (Male)

toon commander who had been captured by the Germans.

This raid is interesting in that it showed the possibility of manoeuvring tanks in the dark through the enemy's lines, and also the great security afforded to the tank by the darkness.

The Battle of Hamel, which was fought on July 4, was the first occasion upon which the Mark V machine went into action. Much was expected of it, and it more than justified all expectations. The object of the attack was a twofold one—firstly, to nip off a salient between the River Somme and the Villers-Bretonneux—Warfusée road; secondly, to restore the confidence of the Australian Corps in tanks, a confidence which had been badly shaken by the Bullecourt reverse in 1917.

As soon as the attack was decided on the training of the Australians with the tanks was commenced at Vaux en Amienois, the headquarters of the 5th Tank Brigade. Tank units for this purpose were affiliated to Australian units and by this means a close comradeship was cultivated.

The general plan of operations was for the 5th Tank Brigade to support the advance of the 4th Australian Division in the attack against the Hamel spur running from the main Villers-Bretonneux plateau to the River Somme. The frontage was about 5,500 yards, extending to 7,500 yards on the final objective, the depth of which was 2,500 yards. The main tactical features in the area were Vaire wood, Hamel wood, Pear-Shape trench, and Hamel village. There was no defined system of trenches to attack except the old British line just east of Hamel, which had been originally sited to obtain observation eastwards. The remainder of the area was held by means of machine-gun nests.

Five companies of 60 tanks in all were employed in the attack; these were divided into two waves—a first-line wave of 48 and a reserve wave of 12 machines. Their distribution was as follows:

6th A.I. Brigade	2 Sections (6 tanks).
4th A.I. Brigade	1 Company (12 tanks).
11th A.I. Brigade	6 Sections (18 tanks).
Liaison between 4th & 11th Brigades	1 Company (12 tanks).
Reserves	1 Company (12 tanks).

The co-operation of the artillery was divided under the headings of a rolling barrage and the production of smoke screens. Behind the former the infantry were to advance followed by the tanks, which were only to pass ahead of them when resistance was encountered. This arrangement was not a good one and was an inheritance of the Bullecourt distrust. The latter were to be formed on the high ground west of Warfusée-Abancourt and north of the Somme and south of Morlancourt. Once the final objective was gained a standing barrage was to be formed to cover consolidation.

As the entire operation was of a very limited character an extensive system of supply dumps was not necessary, so instead each fighting tank carried forward ammunition and water for the infantry and four supply tanks were detailed to carry R.E. material and other stores. Each of these eventually delivered a load of about 12,500 lb. within 500 yards of the final objective, and within half an hour of its capture. The total load delivered on July 4, at 40 lb. per man, represented the loads of a carrying party 1,250 men strong. The number of men used in these supply tanks was twenty-four.

The tanks assembled at the villages of Hamelet and Fouilloy on the night of July 2–3, without hostile interference. On the following night they moved forward to a line approximately 1,000 yards west of the infantry starting-line under cover of aeroplanes which, flying over the enemy, drowned the noise of the tank engines.

Zero hour was fixed for 3.10 a.m. and tanks were timed to leave their starting-line at 3.2 a.m. under cover of artillery harassing fire, which had been carried out on previous mornings in order to accustom the enemy to it. This fire lasted seven minutes, then a pause of one minute occurred, to be followed by barrage fire on the enemy's front line for four minutes. This allowed twelve minutes for the tanks to advance an average distance of 1,200 yards before reaching the infantry line at zero plus four minutes, when the barrage was to lift. All tanks were on the starting-line up to time, which is a compliment to the increased reliability of the Mark V machine over all previous types.

As the barrage lifted the infantry and tanks moved forward. The position of the tanks in relation to the infantry varied, but generally speaking, the tanks were in front of the infantry and immediately behind the bursting shells. The enemy's machine-gunners fought tenaciously, and in several cases either held up the infantry, or would have inflicted severe casualties on them, if tanks had not been there to destroy them. The manoeuvring power of the Mark V tank was clearly demonstrated in all cases where the infantry were held up by machine guns, it enabled the tanks to drive over the gunners before they could get away. There were a great number of cases in which the German machine guns were run over and their detachments crushed. Driving over machine-gun emplacements was the feature of this attack; it eliminated all chance of the enemy "coming to life" again after

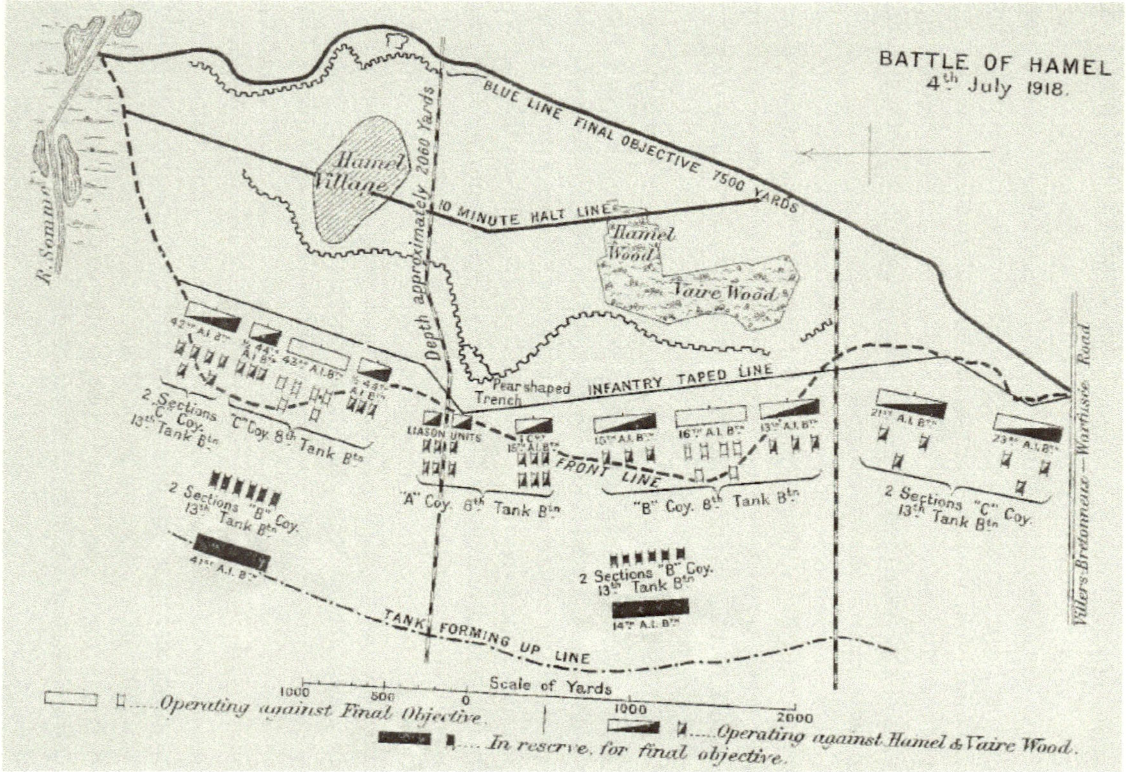
BATTLE OF HAMEL
4th July 1918.
BLUE LINE FINAL OBJECTIVE 7500 YARDS
10 MINUTE HALT LINE
Hamel Village
Hamel Wood
Vaire Wood
Depth approximately 2060 Yards
Pear shaped Trench
INFANTRY TAPED LINE
FRONT LINE
TANK FORMING UP LINE
R. Somme
Villers-Bretonneux—Warfusée Road
42nd A.I. Bn
½ 44th A.I. Bn
43rd A.I. Bn
½ 44th A.I. Bn
2 Sections "C" Coy. 13th Tank Bn
"C" Coy. 8th Tank Bn
LIASON UNITS
½ C Coy 15th A.I. Bn
10th A.I. Bn
16th A.I. Bn
13th A.I. Bn
21st A.I. Bn
23rd A.I. Bn
"A" Coy. 8th Tank Bn
"B" Coy. 8th Tank Bn
2 Sections "C" Coy. 13th Tank Bn
2 Sections "B" Coy. 13th Tank Bn
41st A.I. Bn
2 Sections "B" Coy. 13th Tank Bn
14th A.I. Bn
Scale of Yards
1000 500 0 1000 2000
....Operating against Final Objective.
....Operating against Hamel & Vaire Wood.
.... In reserve, for final objective.

the attack had passed by.

Tanks detailed for the right flank had severe fighting and did great execution, their action being of the greatest service. The tanks detailed to support the infantry battalions passing round Vaire and Hamel woods and Hamel village guarded their flanks whilst this manoeuvre was in operation.

The tank attack came as a great surprise to the Germans and all objectives were taken up to scheduled time. The enemy suffered heavy loss, and, besides those killed, 1,500 prisoners were captured. The 4th Australian Division had 672 officers and other ranks killed and wounded, and the 5th Brigade had only 16 men wounded and 5 machines hit. These tanks were all salved by the night of July 6–7.

The co-operation between the infantry and tanks was as near perfect as it could be; all ranks of the tank crews operating were impressed by the superb moral of the Australian troops, who never considered that the presence of the tanks exonerated them from fighting, and who took instant advantage of any opportunity created by the tanks. From this day on the fastest comradeship has existed between the Tank and Australian Corps. Bullecourt was forgotten, and from the psychological point of view this was an important objective to have gained prior to the great attack in August.

The second battle in which the Tank Corps took part in July was the Battle of Moreuil or Sauvillers; it is of particular interest, for it was the only occasion during the war in which our tanks, in any numbers, operated with the French Army. Another interesting point connected with this attack was the rapidity with which it was mounted.

At 2.30 p.m. on July 17 the 5th Tank Brigade Commander was informed that he was to prepare forthwith to co-operate in an attack to be made by the IXth Corps of the First French Army, and, for this purpose, the 9th Battalion of the 3rd Tank Brigade was to be placed under his command and that he and this battalion would come under the orders of the 3rd French Infantry Division.

The object of the operation was a threefold one:

(1) To seize the St. Ribert wood with the object of outflanking Mailly-Raineval from the south.

(2) To capture the German batteries in the neighbourhood of St. Ribert or to force them to withdraw.

(3) To advance the French field batteries eastwards in order to bring fire to bear on the ridge which dominates the right bank of the River Avre.

BATTLE OF MOREUIL OR SAUVILLERS
July 23rd 1918.

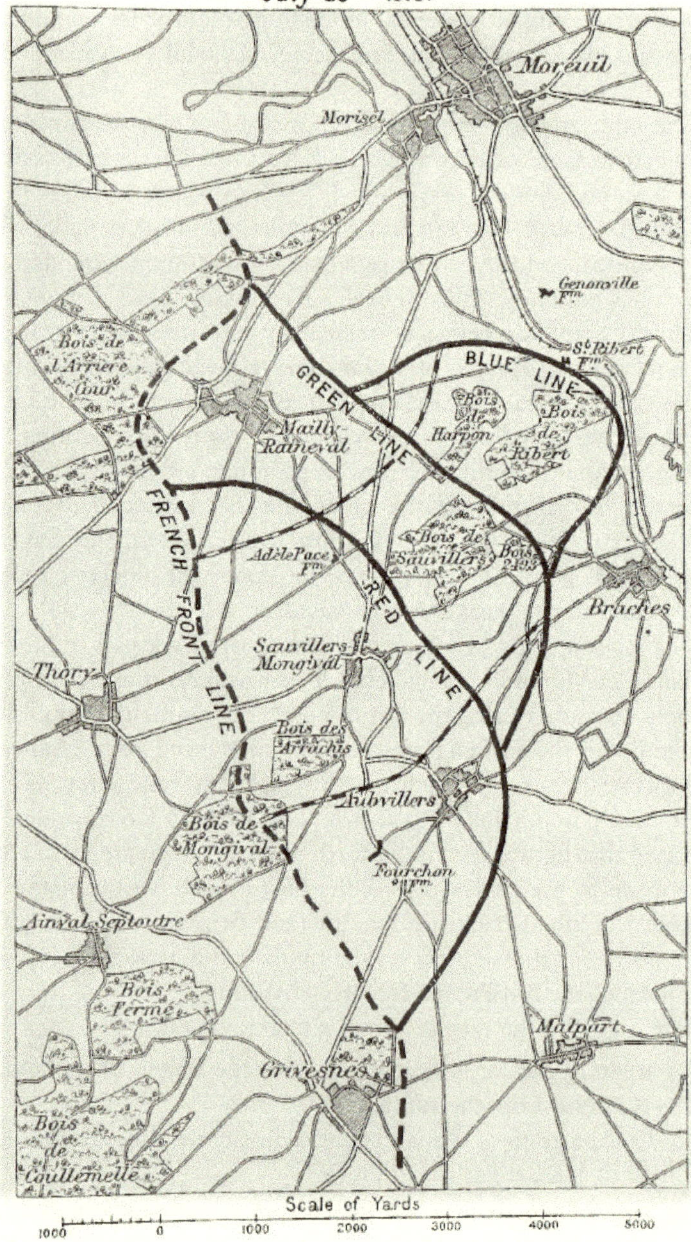

The three French Divisions attacking were the 152nd Division on the right, the 3rd in the centre, and the 15th on the left. Their respective frontages were 950, 2,000, and 800 metres. The greatest depth of the attack was 3,000 metres.

The operation was to be launched as a surprise, after a short and intense artillery preparation; the main objectives were to be captured by encircling them and then "mopping" them up.

There were three objectives: The first included Bois des Arrachis, Sauvillers, Mongival village, Adelpare farm and Ouvrage-des-Trois-Bouqueteaux; twelve tanks and four battalions of infantry were detailed for this. The second included the clearing of the plateau to the north of the Bois-de-Sauvillers and the capture of the south-west corner of the Bois-de-Harpon; the number of tanks allotted to this objective was twenty-four, with four fresh infantry battalions. The third, known as the Blue Line, an outpost line covering the second objective, was to be occupied by eight strong infantry patrols and all available machines.

The attack was to be preceded by one hour's intensive bombardment, including heavy counter-battery fire. The creeping barrage was to consist of H.E. and smoke shells and was to move at the rate of 200 metres in six minutes up to the first objective, after this at the rate of 200 metres in eight minutes.

Tanks were to attack in sections of three, two in front and one in immediate support, the infantry advancing in small assaulting groups close behind the tanks.

Directly the orders were issued preparations were set on foot. On July 18, Lieutenant-Colonel H. K. Woods, commanding the 9th Battalion, and his reconnaissance officers visited General de Bourgon, the Commander of the 3rd French Division, who explained to them the scheme; on the next day these officers reconnoitred the ground over which the battalion would have to operate, and tactical training was carried out with the French at the 5th Tank Brigade Driving School at Vaux-en-Amienois. (General de Bourgon was a great friend of the Tank Corps; he presented its Headquarters mess with a charming trophy.) On the 20th and 21st training continued, and further examination of the ground was made, and on the 22nd details of the attack were finally settled. In spite of the continuous exertion of the last few days all ranks were in the greatest heart to show the 3rd French Division what the British Tank Corps could do.

Meanwhile headquarters were selected, communications arranged for, supplies dumped, and reorganisation and rallying-points worked out and fixed.

The move of the 9th Battalion is particularly interesting on account of its rapidity. On July 17 it was in the Bus-les-Artois area; on the 18th it moved 16,000 yards across country and entrained under sealed orders at Rosel, detraining at Conty. On the night of the 19th–20th it moved 4,000 yards from Conty to Bois-de-Quemetot; on 20th–21st, 9,000 yards to Bois-de-Rampont; on 21st–22nd, 7,000 yards to Bois-de-Hure and Bois-du-Fay; and on 22nd–23rd 4,500 yards from these woods into action with thirty-five machines out of the original forty-two fit to fight.

The country over which the action was to be fought was undulating, and with the exception of large woods there were few tank obstacles. Prior to the operations the weather had been fine, but on the day of the attack heavy rain fell and visibility was poor, a south wind of moderate strength was blowing.

The preliminary bombardment began at 4.30 a.m., and, an hour later, the tanks having been moved up to their starting-points without incident, the attack was launched. The tanks advanced ahead of the infantry, Arrachis wood was cleared and Sauvillers village attacked, the tanks occupying this village some fifteen minutes before the infantry arrived. At Adelpare farm and Les-Trois-Bouqueteaux the enemy's resistance, as far as the tanks were concerned, was light, and the German machine-gun posts were speedily overrun. From Sauvillers village, at zero plus two hours, the tanks advanced on to Sauvillers wood, which, being too thick to enter, had to be skirted, broadsides being fired into the foliage. Whilst this was proceeding other tanks moved forward towards the Bois-de-St.-Ribert, but as the infantry patrols did not appear, they turned back to regain touch with the French infantry. About 9.30 a.m., whilst cruising round, six tanks were put out of action in rapid succession by direct hits fired from a battery situated to the south of St. Ribert wood.

At 9.15 a.m. an attack on Harpon wood was hastily improvised between the O.C. B Company, 9th Battalion, and the commander of one of the battalions of the 51st Regiment. This attack was eminently successful; the French infantry, following the tanks with great *élan*, established posts in Harpon wood. After this action the tanks rallied.

In this attack the tank casualties were heavy in personnel: 11 officers and men were killed and 43 wounded, and 15 tanks were put out of action by direct hits. The losses in the French Divisions were: 3rd—26 officers and 680 men; 15th—15 officers and 500 men; 152nd—20 officers and 650 men. It should be noted that though the 3rd, with which tanks co-operated, had to attack the largest system of defences, its casualties approximately equalled those of each of the other divisions.

The number of prisoners captured was 1,858, also 5 guns, 45 trench mortars, and 275 machine guns.

After the attack, when the tanks had returned to their positions of assembly, General Debeney, commanding the First French Army, paid the 9th Battalion the great honour of personally inspecting it on July 25, and of expressing his extreme satisfaction at the way in which the Battalion had fought. As a token of the fast comradeship which had now been established between the French troops of the 3rd Division and the 9th Tank Battalion, this battalion was presented with the badge of the 3rd French Division and ever since this day the men of this unit have worn it on their left arm.

CHAPTER 28
German Tank Operations

In spite of the fact that throughout the war the Germans never had at their disposal more than some fifteen tanks of their own manufacture and some twenty-five captured and repaired British Mark IV machines, their employment of these machines is worth recording.

As already mentioned the Germans learnt little from the Mark I machine they captured and held for several days during the Battle of the Somme. In fact, they appear to have treated the tanks generally, during these operations, with scorn. The machine was indeed mechanically indifferent, but the German, who is essentially a stupid (*dumm*) man, could not apparently differentiate between the defects of mechanical detail and the advantages of fundamental principles, such as mobility, security, and offensive power, which indeed the whole "idea" of the tank represented.

The action at Bullecourt, it is thought, opened the German eyes to the possibilities of a tank attack, that is an attack in which tanks are used as the resistance-breakers in advance of the infantry. If two tanks could accomplish what the two Mark I's did on April 11, 1917, there was no reason why 200 should not win a great victory, and 2,000 end the war. Be this as it may, it was at about this time—the spring of 1917—that the first German tank construction was begun at the Daimler works near Berlin, and the result of this was the production of fifteen machines known as "Type A.7.V." (see Plate 6), some of which first took the field in March 1918.

The chief characteristics of this tank were: its good speed on smooth ground, on which it could attain some eight miles an hour; its inability to cross almost any type of trench or shelled ground on account of its shape. In weight it was about 40 tons, it carried very thick armour especially in front, capable of withstanding A.P. bullets at close range and field-gun shells, not firing A.P. ammunition, at long; it was, however, very vulnerable to the splash of ordinary bullets on account of the crevices and joints in its armour. The most interesting feature of this otherwise indifferent machine was that its tracks were provided with sprung bogies. The use of sprung tracks in so heavy a tank was the only progressive step shown in the German effort at tank production.

The German tank was 24 ft. long and 10 ft. 6 in. wide; its armament was one 1·57 mm. gun and 6 machine guns; its crew, one officer, eleven N.C.O.s, and four private soldiers—exactly twice the strength of the crew of a British Mark IV tank. This crew comprised three distinct classes, drivers (mechanics), gunners (artillerymen), and machine-gunners (infantrymen). These three classes remained distinct, little co-operation existing between them.

Both the tanks of German manufacture and the captured British tanks were divided into sections (Abteilungen) of five machines each, the personnel establishment of which was as follows:

	German Tanks	*Captured Tanks*
Captain Commanding	1	1
Lieuts. or 2nd Lieuts.	5	5
Drivers	81	81
Machine-gunners	48	20
Artillerymen	22	14
Signallers	12	12
Medical Corps	1	1
Orderlies, etc.	6	6
Total	176 All ranks	140 All ranks

This establishment was a very extravagant one when compared with that of a Mark IV section of five tanks, namely, six officers and thirty-five other ranks.

Besides the "A.7.V." machines the Germans employed, during their various offensives of 1918, a number of caterpillar ammunition carriers known as "*Munitions Schlepper*," or "*Tankautos*." These could proceed across country as well as by road.

The morale of the German Tank Corps was not high, and as regards the personnel of the captured Mark IV Tanks it was decidedly low, the Germans having made considerable efforts to prove to their own troops, by means of demonstrations, that this type of tank was both vulnerable and ineffective. The training of this Corps appears to have been indifferent; a certain number of Assault Divisions were trained with wagons representing tanks, and in a few cases, it is believed that actual tanks were used with infantry in combined training.

The tactics of the German tanks simply consisted in the "mopping up" of strong points. On several occasions they did get in front of the attacking infantry, but they do not appear in any sense to have led the attack. The following extract from the German G.H.Q. instructions, "The Co-operation of Infantry with Tanks" (!), indicates that no real co-operation was ever contemplated. It reads:

> The infantry and tanks will advance independently of one another. No special instructions regarding the co-operation with tanks will be issued. When advancing with tanks the infantry will not come within 160 yards of them on account of the shells which will be fired at the tanks.

In all, there are nine recorded occasions upon which the Germans made use of tanks, the first of which was in their great offensive which opened on March 21, 1918. In this attack about ten German and ten captured British machines were used, and although they accomplished very little, they were much written up in the German press.

A little over a month later, on April 24, the only successful German tank attack during the war was carried out. On this occasion, which has been referred to in Chapter 26, fourteen tanks were brought forward, and of these twelve came into action and captured Villers-Bretonneux, a point of great tactical importance; a counter-attack carried out by the Australian Corps and a few British tanks, however, restored the situation.

A month later a few tanks were used by the Germans against the French on the opening day of the great Aisne offensive, namely May 27. None of these machines, however, succeeded in passing a large trench in the second defensive system known as Dardanelles trench.

On June 1, fifteen operated with little success in the Reims sector, eight being left derelict in the French lines. Similar unsuccessful

PLATE 6

French Renault Tank.

German Tank

operations were carried out on June 9 and July 15.

On August 31, three German tanks approached our lines east of Bapaume; two were knocked out by our guns and eventually captured.

On October 8, some fifteen captured British machines were used against us in the Cambrai sector. Of this action the German account states that these tanks were employed defensively to fill up a gap in their line; whether this was so or not, they undoubtedly produced a demoralising effect amongst our own men, equilibrium only being re-established when two of them were put out of action. Three days later, on the 11th, a few tanks were used at St. Aubert; this was the last recorded occasion upon which the Germans made use of tanks in the Great War.

Indifferent as were the German tank tactics as compared with our own, one fact was most striking, this being that the British infantry no more than the German would or could withstand a tank attack. The reason for this is a simple one, namely, inability to do so. So pronounced was this feeling of helplessness that when, during our own retirement in March 1918, rumours were afloat that German tanks were approaching, our men in several sectors of the line broke and fell back. During the German retirement a few months later on we find exactly the same lowering of morale by self-suggested fear, fear based on the inability to overcome the danger. This moral effect produced by the tank was appreciated by the Germans, for in a note issued by the XVIIth German Army we find:

> Our own tanks strengthen the morale of the infantry to a tremendous extent, even if employed only in small numbers, and experience has shown that they have a considerable demoralising effect on the hostile infantry.

CHAPTER 29
The Battle of Amiens

On July 15 the renewed German offensive on the Château-Thierry—Reims front had been launched and failed. Strategically and tactically placed in as unenviable a position as any army well could be, the Crown Prince's forces received a staggering blow on the 18th, when Marshal Foch launched his great tank counter-attack against the western flank of the Soissons salient.

At the time of this attack the brigades of the Tank Corps were distributed defensively along the First, Third, and Fourth Army fronts, in order to meet by counter-attacks any renewal of the enemy's offensive against these armies.

Ever since the dramatic *coup-de-main* accomplished on July 4 by the 4th Australian Division and the 5th Tank Brigade in the Battle of Hamel, the general interest in tanks had become much more conspicuous. The great tank attack at the Battle of Cambrai, convincing in worth as it was to all who had taken part in it, had been somewhat discredited by the recent German offensive on the Somme front, which was seized upon by certain soldiers of the old school to reinforce their assertion—that the day of the tank had come and gone, and that to fight a second Battle of Cambrai was too great a gamble to be worth risking. Now a series of projects were asked for which embraced various areas of operation; in the Fourth Army against the Amiens salient; in the Third Army against Bucquoy and Bapaume; in the First Army against the Merville salient and in the Second against Kemmel hill. The only one of these projects which offered prospects of a decisive success was the first.

On July 13 the Fourth Army Commander was asked by G.H.Q. to submit a scheme for an attack on his front. This was done on the 17th, when a limited operation, with the object of capturing the Amiens outer defence line, running from Castel through Caix to Mericourt, was outlined. The force suggested for this attack was three corps and eight battalions of tanks. On the 21st a conference was held at the Fourth Army headquarters at Flixecourt when, on the suggestion of the Tank Corps, the number of tank battalions was raised from eight to twelve; this comprised the whole Tank Corps less the 1st Tank Brigade, which was still armed with Mark IV machines, and which at this time was engaged in training its personnel on the Mark V tank.

On July 27, zero day was fixed for the 10th, but on August 6 this was changed to the 8th. All this time, in order to maintain secrecy, no mention of the impending attack was permitted, and the only preparation which could be undertaken was to send one officer of the Tank Corps General Staff to the area of operations to study the ground. On July 30 a conference was held at the 5th Tank Brigade headquarters at Vaux, at which the Fourth Army Commander explained the plan of operations. From this day on preparations were begun, the railway moves being issued the same evening.

As already stated, the original proposal was a limited operation, the centre of the attack being carried out by the Canadian and Australian Corps. The right of the Canadians was to be covered by the French

First Army attacking east and south-east of the Luce River. The left of the Australians was to be protected by two divisions of the IIIrd Corps operating towards Bray. On July 29 the scope of the operation was extended as follows:

To disengage the Amiens-Paris railway by occupying the line Hangest—Harbonnières—Mericourt.

To advance to the line Roye—Chaulnes, driving the enemy towards Ham, and so facilitate the advance of the French on the line Noyon-Montdidier.

The force placed at the disposal of the Fourth Army consisted of the following Corps:

(1) The Canadian Corps—4 divisions.

(2) The Australian Corps—4 divisions.

(3) The IIIrd Corps—2 divisions.

(4) General Reserve—3 divisions, to be supplemented by further divisions as soon as possible.

(5) The Cavalry Corps—3 cavalry divisions.

Tank battalions were allotted to the 3 infantry corps as follows:

(1) Canadian Corps, 4th Tank Brigade—1st, 4th, 5th, and 14th Battalions.

(2) Australian Corps—5th Tank Brigade—2nd, 8th, 13th, and 15th Battalions.

(3) IIIrd Corps—10th Battalion.

(4) General Reserve—9th Battalion (still refitting at Cavillon).

(5) Cavalry Corps. 3rd Tank Brigade—3rd and 6th Battalions.

The 3rd and 6th Battalions were equipped with 48 Whippet tanks each; all the other battalions were heavy units equipped with 42 Mark V machines each (36 fighting and 6 training tanks), except the 1st and 15th Battalions, which were each equipped with 36 Mark V One Star machines.

As in the Battle of Cambrai the initiation of the attack was to depend on the tanks, no artillery registration or bombardment being permitted prior to the assault. In all, some 82 brigades of field artillery, 26 brigades of heavy artillery, and 13 batteries of heavy guns and howitzers were to be employed. The following is a summary of the artillery instructions:

(1) No artillery bombardment.

(2) The initial attack to be opened by a barrage at zero.

(3) The majority of the heavy guns and howitzers to concentrate on counter-battery work.

(4) The field-artillery brigades to be prepared to move forward and offer the closest support to the attacking infantry.

(5) Special noise barrages to cover the approach of the tanks.

The first object of the Cavalry Corps was to secure the old Amiens defence line and hold it until relieved by infantry units. The second, to push forward on the line Roye—Chaulnes. For this purpose, the 3rd Cavalry Division with one battalion of Whippets was placed under the command of the Canadian Corps, and one cavalry brigade, supported by one company of Whippets, under that of the Australian Corps.

On July 30, the date on which preparations were begun, the Tank Corps was distributed as follows:

1st Tank Brigade	7th Battalion	Merlimont.
H.Q. Estruvalle	11th ,,	Merlimont.
	12th ,,	Merlimont.
2nd Tank Brigade	10th ,,	Bouvigny.
H.Q. Bois D'Ohlain	14th ,,	Mont St. Eloi.
	15th ,,	Simencourt.
3rd Tank Brigade	3rd ,,	Toutencourt.
H.Q. Wavrans	6th ,,	Merlimont.
4th Tank Brigade	1st ,,	Coullemont.
H.Q. Couturelle	4th ,,	La Cauchie.
	5th ,,	Bailleulval.
5th Tank Brigade	2nd ,,	Querrieu wood.
H.Q. Vaux	8th ,,	Blangy (east of Amiens).
	13th ,,	St. Gratien (near to).
	9th ,,	Cavillon.

In order to facilitate co-operation and staff work it was decided to break up temporarily the 2nd Tank Brigade and to allot the 10th Battalion to the IIIrd Corps and the 14th and 15th Battalions to the 4th and 5th Tank Brigades respectively. Besides these units five Supply and Gun Carrier Companies were allotted for the transport of tank and infantry supplies.

Briefly the general preparations were carried out as follows: the 1st, 4th, 5th, 10th, 14th, and 15th Battalions were concentrated by rail in the Fourth Army area, detraining at Poulainville, Saleux, Prouzel, and Vignacourt between July 31 and August 5. The 3rd and 6th Whippet Battalions moved to Naours by the night of August 2–3, and thence to the Boulevard Pont-Noyelles in Amiens on the night of the 6th–7th, where they lay hid under the trees. Tanks were got ready commencing on July 31, on which date the formation of supply dumps was begun.

Plate 7

Gun-Carrier

Mark V Star Tank (Female)

The 9th Battalion, which had been withdrawn to Cavillon after the Battle of Moreuil, was allotted to the Canadian Corps for training; the training of the Australian Corps continuing as heretofore at Vaux. Considering the short time available for preparation the speed with which this great battle was mounted redounds to the credit of all ranks taking part in it. It was a triumph of good staff work.

The detailed preparations of the four groups of tanks—3rd, 4th, and 5th Brigades, and the 10th Battalion, are interesting and were as follows:

Fourth Tank Brigade.—This brigade established its advanced headquarters at Dury. Its battalions were distributed as follows: the 1st, 4th, 5th, and 15th Battalions to the 4th, 1st, 3rd, and 2nd Canadian Divisions respectively.

No. 3 Tank Supply Company was split up amongst the divisions of the Canadian Corps; three forward wireless stations were arranged for as well as one back receiving station; assembly positions and rallying-points were fixed, and the 2nd Tank Field Company was detailed, once the battle began, to clear all obstacles off the Berteaucourt-Thennes road and to prepare crossings over the Luce River between Hangard and Demuin.

The plan of the Canadian Corps attack was as follows:

The 1st, 2nd, and 3rd Canadian Divisions were to make good the Red Line on zero day, except the left of the 2nd Division, which was to push on and occupy the Blue Line. The advance of the 3rd Canadian Division was timed to start at zero plus four hours, the time it was considered the initial attack would leave the Green Line. The 4th Canadian Division was to follow the 1st and 3rd to the Blue Line and then to the line Moreuil—Demuin—Marcelcave. The 1st Tank Battalion was allotted to this division, and arrangements were made for each of its tanks to carry forward two Lewis and two Vickers gun-teams besides the crew, these units being intended to assist the cavalry on the Blue Dotted Line. Besides the above an independent force, consisting chiefly of Canadian motor machine guns, was to operate down the Roye road.

Fifth Tank Brigade.—The 5th Tank Brigade established its advanced headquarters at Hospice Fouilloy with a report centre at the northwest corner of Kate wood. Its battalions were distributed as follows: the 2nd Battalion and one company of the 13th Battalion to the 2nd and 5th Australian Divisions; the 13th Battalion less one company to the 3rd Australian Division; the 8th Battalion to the 4th; the 15th Battalion was split into halves of eighteen tanks each, one half operating with the 4th and the other with the 5th Australian Divisions.

As regards supply arrangements, No. 1 Gun Carrier Company was allotted to the Australian Corps for transport work. Two forward wireless stations and one back receiving station were fixed, and assembly and rallying-points settled.

The general plan was that the tanks were to advance to the first objective under an artillery barrage. On reaching the second objective all tanks were to rally except those of the 15th Battalion, which were to push on to the Blue Dotted Line carrying machine-gunners forward.

Tenth Battalion.—The whole of the tanks of the 10th Battalion, less one section, were to operate against the first objective and then push on to the second, after which they were to rally west of the first objective.

Third Tank Brigade.—The 3rd Tank Battalion was allotted to the 3rd Cavalry Division and the 6th Tank Battalion to the 1st Cavalry Division. The objective of these two battalions was to secure the area between the Red Line and the old Amiens defence line. The advance of the 3rd and 1st Cavalry Divisions was to take place at zero plus four hours. Before the Red Line was reached the cavalry scouts were to precede the Whippet tanks and discover crossings over the Luce River at Ignaucourt and Demuin. If crossings were found the Whippets were to use them; otherwise they were to advance eastwards near Caix. The formation to be adopted by these two battalions was one company to act as a screen in front of the cavalry with 200 yards interval between tanks, one company in support, and one company in reserve.

The country between our front line and the line Roye—Frise was in every respect suitable to tank movement. East of the Roye—Frise line began the French portion of the old Somme battlefield; the ground here in places had been heavily shelled, but was quite negotiable by heavy tanks. The flanks of the attack were the two difficult points. Neither permitted of the use of offensive wings and both offered good defensive positions for the enemy's machine-gunners.

Zero was at 4.45 a.m., when 415 fighting tanks out of 420 went into action; this in itself was a notable feat of mechanical efficiency.

★★★★★★

Nine heavy battalions with 324 machines and two medium

BATTLE OF AMIENS
August 8th 1918.

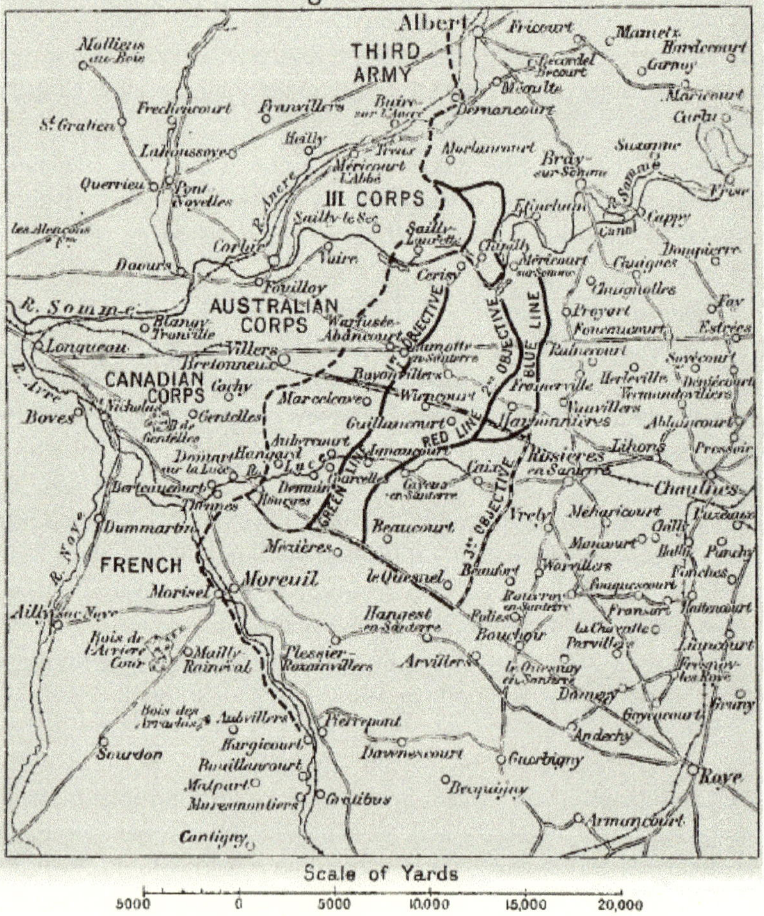

battalions with 96. Besides these tanks, there were 42 in mechanical reserve, 96 supply tanks, and 22 gun-carriers. In all, and not counting the machines of the 9th Tank Battalion, there were 580 tanks.

★★★★★★

The attack was an overwhelming surprise and, though the enemy was holding his line in strength, little opposition was met with except in a few localities. At the Battles of Hamel and Moreuil the German machine-gunners had learnt to appreciate what the increased mobility of the Mark V tank enabled it to accomplish, and not being anxious to be crushed under its thirty tons of steel they gave less trouble during this battle than on any previous occasion. In spite of this, many hostile machine-gun posts were hunted out of the standing corn and run over. Co-operation was throughout good, especially on the Canadian and Australian fronts, where the attack swept on irresistibly. On the IIIrd Corps front the attack started in a state of some confusion, due to the fog and the uncertain state of the line, the Germans having attacked the IIIrd Corps on the 6th, and the IIIrd Corps having retaken most of their lost trenches on the 7th. This undoubtedly complicated the attack on the 8th.

South of the River Somme all objectives were taken up to time; on the right flank the difficult valley of the Luce was crossed by all except two tanks; this was a high compliment to the crews working on this flank, for the approach of the tanks was rendered most difficult on account of fog.

Both battalions of Whippets were engaged with their respective cavalry divisions and had a considerable amount of fighting to do in the neighbourhood of Cayeux wood, Le Quesnel, east of Mezières, at Guillaucourt and the railway south of Harbonnières, which was held with great determination by the enemy as far as the Rosières-Vauvillers road.

During this day's fighting a total of 100 machines were temporarily put out of action chiefly by the enemy's fire from the Chipilly ridge, which, on account of the partial failure of the IIIrd Corps attack, was held by the enemy for several days after August 8. On the evening of the 8th the tanks rallied; the crews, however, were so exhausted by the great distance covered, the maximum penetration effected being about 7½ miles, and the heat of the day, that it was necessary to resort to the formation of composite companies for the next day's operation, few reserves remaining in hand, and the 9th Tank Battalion, which was now moving eastwards from Cavillon, was not in a position to take the field for at least forty-eight hours.

On the night of August 8–9, the front line of our attack from north to south ran approximately as follows: along the outer Amiens defence line to Proyart—west of Rainecourt—east of Vauvillers—east of Rosières—east of Meharicourt—east of Rouvroy—east of Bouchoir. South of the Amiens-Roye road the line was continued by the French, who had captured Hangest, Arvillers, and Pierrepont.

Up to 6 o'clock on the morning of August 9, some 16,000 prisoners had passed through the British and French cages, and over 200 guns had been counted. Many prisoners testified to the rapid advance of the tanks which, appearing suddenly out of the mist, rendered all resistance useless. It is interesting to record that those prisoners who had seen tanks before all noticed that they were up against a new type which moved faster and manoeuvred better than the old ones.

On the evening of the 8th orders were issued that the attack should be resumed on the following morning with a view to advancing it to the line Roye—Chaulnes—Bray-sur-Somme—Dernancourt, particular attention being paid to the left flank. A strong position was to be established north of the Somme in order to form a defensive flank to the Fourth Army.

On August 9, north of the Somme, the 10th Battalion put sixteen tanks into action with the 12th and 58th Divisions. The attack was, however, at first held up by machine-gun fire from the woods round Chipilly, and the work entailed in engaging these weapons by means of tanks was found most difficult on account of the steep valleys in this sector and the close nature of the woods. Later on in the day, objectives were gained, but only after five tanks had been put out of action.

South of the Somme the 5th and 4th Tank Brigades attacked the front Framerville-Rosières-Bouchoir with 89 tanks. Near Lihons five machines received direct hits, but in the action round Framerville out of the 13 tanks engaged only 1 was hit. The fewness of tank casualties here was undoubtedly due to the excellent infantry co-operation, riflemen working hand-in-hand with the tanks and picking off the enemy's gunners directly the machines came under hostile artillery observation.

The 3rd Tank Brigade's action with the Cavalry Corps was disappointing, the tanks being kept too long at their Brigade Headquarters. At Beaufort and Warvillers the Whippets rendered great assistance to the infantry by chasing hostile machine-gunners out of the crops and shooting them down as they fled.

On this day in all 145 tanks went into action, of which 39 were hit by hostile gun fire.

On the night of August 9–10, the attack had reached the line Bouchoir—Warvillers—Rosières—Framerville—Mericourt. On the 10th the Fourth Army orders were to continue the advance with the object of gaining the general line Roye—Chaulnes—Bray-sur-Somme—Dernancourt. New French forces were also going to attack on the front south of Montdidier.

On the morning of August 10, the 10th Battalion co-operated in two small attacks carried out by the 12th Division. Seven tanks took part and attacked the enemy north of Morlancourt and along the Bray-Corbie road. This was the last action fought on this front by this battalion.

South of this, the 5th Tank Brigade carried out a minor night operation against Proyart, and the 4th Tank Brigade with 43 tanks supported the 32nd Division, fresh from the general reserve, and the 4th Canadian Division in an attack on the line Roye—Hattencourt—Hallu; owing to the late issue of orders, the hour of attack was altered, and eventually the advance took place in daylight without smoke. A stubborn resistance was met with, and out of the 43 tanks operating no fewer than 23 received direct hits.

The Whippets with the cavalry fared equally badly on this day. They were ordered to capture Parvillers, but neither the cavalry nor Whippets reached this spot owing to the old trench systems and the broken nature of the ground. The edge of the old Somme battlefield had now been reached, and the time was rapidly approaching when the shelled area would offer as great an obstacle to the attack as it would an assistance to the retiring enemy.

During the 10th some 67 tanks in all were engaged, and of these 30 received direct hits.

On August 11 no appreciable change took place on the British front. Lihons was, however, captured by the 1st Australian Division, assisted by ten tanks of the 2nd Battalion, otherwise most of the tank operations consisted in mopping up strong points. On the evening of this day the 4th and 5th Tank Brigades were withdrawn from action to refit.

During the next few days it was decided that, whilst pressure should be kept up south of the Somme, a new battle should open to the north of this river on the Third Army front, and that three Tank Brigades should co-operate in this attack; this necessitated the transfer of the 4th Tank Brigade to the IIIrd Corps north of the Somme and the withdrawal of the 10th, 14th, and 15th Battalions from the Fourth Army area; this left the 4th Tank Brigade with the 1st, 4th, and 5th Battalions, and the 5th Tank Brigade with the 2nd, 8th, and 13th Battalions.

On August 17 the general situation was as follows: A total of 688 tanks had been in action on August 8, 9, 10, and 11; 480 machines had been handed over to Salvage; very few of the remaining machines were actually fit for a lengthy action, and all required a thorough overhaul; four days, as we shall see, were only possible for this, for the next battle was scheduled to open on August 21.

The great Battle of Amiens was now at an end. A tremendous physical, and above all, moral blow had been dealt the enemy; not only had he lost 22,000 prisoners and 400 guns, but also all hope of winning the war by force of arms. On August 16 the Fourth Army Commander, General Sir Henry Rawlinson, issued the following Special Order, which sums up the reason for this great victory:

> The success of the operations of August 8 and succeeding days was largely due to the conspicuous part played by the 3rd, 4th, and 5th Brigades of the Tank Corps, and I desire to place on record my sincere appreciation of the invaluable services rendered both by the Mark V and the Mark V Star and the Whippets.
>
> The task of secretly assembling so large a number of tanks entailed very hard and continuous work by all concerned for four or five nights previous to the battle.
>
> The tactical handling of the tanks in action made calls on the skill and physical endurance of the detachments which were met with by a gallantry and devotion beyond all praise.
>
> I desire to place on record my appreciation of the splendid success that they achieved, and heartily to congratulate the Tank Corps as a whole on the completeness of their arrangements and the admirable prowess exhibited by all ranks actually engaged on this occasion.
>
> There are many vitally important lessons to be learned from their experiences. These will, I trust, be taken to heart by all concerned and made full use of when next the Tank Corps is called upon to go into battle.
>
> The part played by the tanks and Whippets in the battle of August 8 was in all respects a very fine performance.
>
> The success of the operations may be attributed to—surprise, the

moral effect of the tanks, the high moral of our own infantry, the rapid advance of our guns, and the good roads for supplies.

The main deductions to be drawn from this battle are:

(1) That once preparations are well in progress it is almost impossible to modify them to meet any change in objective.

(2) That the staying power of an attack lies in the general reserve. In this attack the tank general reserve was very weak, consequently after August 8 tank attacks began to "peter out."

(3) That the heavy tank is an assault weapon. Its role is in trench warfare. Once open warfare is entered on infantry must protect tanks from artillery fire.

(4) That the endurance in action of heavy tanks may, at present, be put down as being three days, after which they require overhaul.

(5) That the supply tank is too slow and heavy; a light machine such as a cross-country tractor should replace it.

(6) That at present wireless and aeroplane communications cannot be relied upon; the safest means of communication and the simplest is by galloper.

(7) That the attachment of tanks to cavalry is not a success; for, in this battle, each of these arms in many ways impeded rather than helped the other. During the approach marches the Whippets frequently were reported to have been unable to keep up with the rapid movement of the cavalry; during actual fighting the reverse took place. By noon on August 8, great confusion was developing behind the enemy's lines, by this time the Whippets should have been operating five to ten miles in advance of the infantry, accentuating this demoralisation. As it was, being tied down to support the cavalry, they were a long way behind the infantry advance, the reason being that as cavalry cannot make themselves invisible on the battlefield by throwing themselves flat on the ground as infantry can, they had to retire either to a flank or to the rear to avoid being exterminated by machine-gun fire. Close co-operation between cavalry and tanks being, therefore, practically impossible, both suffered by attempting to accomplish it.

The outstanding lesson of the Battle of Amiens as far as tanks are concerned is that neither the Mark V nor the Whippet machine has sufficient speed for open warfare. Had we possessed a machine which could have moved at an average rate of ten miles an hour, which had a radius of action of 100 or more miles, in this battle we should have not only occupied the bridges across the Somme between Peronne and Ham by noon on August 8, but, by wheeling south-east towards Noyon, we should have cut off the entire German forces south of the Amiens—Roye—Noyon road and inflicted such a blow that in all probability the war would have ended before the month was out. Both from the positive and negative standpoint, this battle may be summed up as "a triumph of machine-power over man-power," or, if preferred, "of petrol over muscle."

CHAPTER 30
The Fight of a Whippet Tank

In this history space has forbidden any extensive reference to individual tank actions, though when all is said and done it was on these actions that not only was the efficiency of the Tank Corps founded but victory itself.

Prior to the Battle of Amiens, it will be remembered that the 3rd and 6th Whippet Battalions were allotted to work with the Cavalry Corps, and that this did not prove a great success owing to the difficulty of combining the action of steel mechanically driven with horseflesh. The account of the action given below is that of a single machine, working well in advance of the attack against the enemy's communications, as it is and was considered at the time in the Tank Corps that all the light tanks should have been. This account is so interesting and instructive that it is quoted in full.

The pluck shown by the crew of one officer and two men, though not exceptional in the Tank Corps, is worthy of the highest praise. These three men, like the Argonauts of old, launched their landship on an expedition faced by unknown dangers; they fought their way through countless odds and faced single-handed the whole of the rear of the German Army. But for an unfortunate accident they might have returned unscathed to safety. In spite of the misfortune which eventually overtook them, at the lowest computation they must have inflicted 200 casualties on the enemy, and at the price of one man killed.

If still there are to be found doubters in the power of the tank, and in the superiority of mechanical warfare over muscular, surely this heroic incident will alone suffice to convince them:

> On August 8, 1918, I commanded Whippet tank 'Musical Box,' belonging to 'B' Company of the 6th Battalion. We left the lying-up point at zero (4.20 a.m.) and proceeded across coun-

try to the south side of the railway at Villers-Bretonneux. We crossed the railway in column of sections, by the bridge on the eastern outskirts of the town. I reached the British front line and passed through the Australian infantry (2nd Australian Division) and some of our heavy tanks (Mark V), in company with the remainder of the Whippets of 'B' Company. Four sections of 'B' Company proceeded parallel with the railway (Amiens—Ham) across country due east. After proceeding about 2,000 yards in this direction, I found myself to be the leading machine, owing to the others having become ditched. To my immediate front I could see more Mark V tanks being followed very closely by Australian infantry.

About this time, we came under direct shell fire from a four-gun field battery, of which I could see the flashes, between Abancourt and Bayonvillers. Two Mark V tanks, 150 yards on my right front, were knocked out. I saw clouds of smoke coming out of these machines, and the crews evacuate them. The infantry following the heavy machines were suffering casualties from this battery. I turned half left and ran diagonally across the front of the battery, at a distance of about 600 yards. Both my guns were able to fire on the battery, in spite of which they got off about eight rounds at me without damage, but sufficiently close to be audible inside the cab, and I could see the flash of each gun as it fired. By this time, I had passed behind a belt of trees running along a roadside. I ran along this belt until level with the battery, when I turned full right and engaged the battery in rear.

On observing our appearance from the belt of trees, the gunners, some thirty in number, abandoned their guns and tried to get away. Gunner Ribbans and I accounted for the whole lot. (This was borne witness to by British troops nearby). I cruised forward, making a detour to the left, and shot a number of the enemy who appeared to be demoralised, and were moving about the country in all directions. This detour brought me back to the railway siding N.N.W. of Guillaucourt. I could now see other Whippets coming up and a few Mark V.s also. The Australian infantry, who followed magnificently, had now passed through the battery position which we had accounted for and were lying in a sunken road about 400 yards past the battery and slightly to the left of it. I got out of my machine and went to an Australian full lieutenant and asked if he wanted any help. Whilst talking to him, he received a bullet which struck the metal shoulder title, a piece of the bullet casing entering his shoulder.

While he was being dressed, Major Rycroft, (Captain of the company to which this tank belonged) on horseback, and Lieutenant Waterhouse, in a tank, and Captain Strachan of 'B' Company, 6th Battalion, arrived and received confirmation from the Australian officer of our having knocked out the field battery. I told Major Rycroft what we had done, and then moved off again at once, as it appeared to be unwise for four machines (Lieutenant Watkins had also arrived) to remain stationary at one spot. I proceeded parallel with the railway embankment in an easterly direction, passing through two cavalry patrols of about twelve men each. The first patrol was receiving casualties from a party of enemy in a field of corn. I dealt with this, killing three or four, the remainder escaping out of sight into the corn. Proceeding further east, I saw the second patrol pursuing six enemy.

The leading horse was so tired that he was not gaining appreciably on the rearmost Hun. Some of the leading fugitives turned about and fired at the cavalryman, when his sword was stretched out and practically touching the back of the last Hun. Horse and rider were brought down on the left of the road. The remainder of the cavalrymen deployed to the right, coming in close under the railway embankment, where they dismounted and came under fire from the enemy, who had now taken up a position on the railway bridge, and were firing over the parapet, inflicting one or two casualties. I ran the machine up until we had a clear view of the bridge, and killed four of the enemy with one long burst, the other two running across the bridge and so down the opposite slope out of sight.

On our left I could see, about three-quarters of a mile away, a train on fire being towed by an engine. I proceeded further east still parallel to the railway, and approached carefully a small valley marked on my map as containing Boche hutments. As I entered the valley (between Bayonvillers and Harbonnières) at right angles, many enemy were visible packing kits and others retiring. On our opening fire on the nearest, many others appeared from huts, making for the end of the valley, their object being to get over the embankment and so out of our sight. We accounted for many of these. I cruised round, Ribbans went into one of the huts and returned, and we counted about sixty dead and wounded. There were evidences of shell fire amongst the huts, but we certainly accounted for most of the casualties counted there. I turned left from the railway and cruised across country, as lines of enemy infantry could be seen retiring. We fired at these many times at ranges of 200 yards to 600 yards. These targets were fleeting, owing to the enemy getting down into the corn when fired on.

In spite of this, many casualties must have been inflicted, as we cruised up and down for at least an hour. I did not see any more of our troops or machines after leaving the cavalry patrols already referred to. During the cruising, being the only machine to get through, we invariably received intense rifle and machine-gun fire. I would here beg to suggest that no petrol be carried on the outside of the machine, as under orders we were carrying nine tins of petrol on the roof, (this was contrary to Tank Corps "Standing Battle Orders.") for refilling purposes when well into the enemy lines (should opportunity occur). The perforated tins allowed the petrol to run all over the cab. These fumes, combined with the intense bullet splash and the great heat after being in action (by this time) nine to ten hours, made it necessary at this point to breathe through the mouthpiece of the box respirator, without actually wearing the mask. At 14.00 hours, or thereabouts, I again proceeded east, parallel to the railway and about 100 yards north of it. I could see a large aerodrome and also an observation balloon at a height of about 200 ft. I could also see great quantities of motor and horse transport moving in all directions. Over the top of another bridge on my left I could see the cover of a lorry coming in my direction. I moved up out of sight and waited until he topped the bridge, when I shot the driver. The lorry ran into a right-hand ditch. The railway had now come out of the cutting in which it had rested all the while, and I could see both sides of it. I could see a long line of men retiring on both sides of the railway, and fired at these at ranges of 400 yards to 500 yards, inflicting heavy casualties. I passed through these and also accounted for one horse and the driver of a two-horse canvas-covered wagon on the far side of the railway.

We now crossed a small road which crossed the main railway, and came in view of a large horse and wagon lines, which ran across the railway and close to it. Gunner Ribbans (right-hand gun) here had a view of the south side of railway, and fired continuously into motor and horse transport moving on three roads (one north and south, one almost parallel to the railway, and one diagonally between these two). I fired many bursts at 600 yards to 800 yards at transport blocking roads on my left, causing great confusion. Rifle and machine-gun fire was not heavy at this time, owing to our sudden appearance, as the roads were all banked up in order to cross the railway. There were about twelve men in the middle aisle of these lines. I fired a long burst at these. Some went down and others got in amongst the wheels and undergrowth. I turned quarter left towards a small copse, where there were more horses and men, about 200 yards away.

On the way across we met the most intense rifle and machine-gun fire imaginable, from all sides. When at all possible we returned the fire, until the left-hand revolver port cover was shot away. I withdrew the forward gun, locked the mounting, and held the body of the gun against the hole. Petrol was still running down the inside of the back door. Fumes and heat combined were very bad. We were still moving forward, and I was shouting to Driver Carney to turn about as it was impossible to continue the action, when two heavy concussions closely followed one another and the cab burst into flames. Carney and Ribbans got to the door and collapsed. I was almost overcome, but managed to get the door open and fell out on to the ground, and was able to drag out the other two men. Burning petrol was running on to the ground where we were lying.

The fresh air revived us and we all got up and made a short rush to get away from the burning petrol. We were all on fire. In this rush Carney was shot in the stomach and killed. We rolled over and over to try to extinguish the flames. I saw numbers of the enemy approaching from all round. The first arrival came for me with a rifle and bayonet. I got hold of this and the point of the bayonet entered my right forearm. The second man struck at my head with the butt end of his rifle, hit my shoulder and neck, and knocked me down. When I came to, there were dozens all round me, and anyone who could reach me did so, and I was well kicked; they were furious. Ribbans and I were taken away and stood by ourselves about twenty yards clear of the crowd. An argument ensued, and we were eventually marched to a dugout where paper bandages were put on our hands. Our faces were left as they were.

We were then marched down the road to the main railway. There we joined a party of about eight enemy, and marched past a field kitchen where I sighed for food. We had had nothing since 8.30 p.m. on the night previous to the action and it was 3.30 p.m. when we were set on fire. We went on to a village where, on my intelligence map, a Divisional H.Q. had been marked. An elderly stout officer interrogated me, asking if I was an officer. I said 'Yes.' He then asked various other questions, to which I replied, 'I do not know.' He said, 'Do you mean you do not know or you will not tell me?' I said, 'You can take it whichever way you wish.' He then struck me in the face and went away. We went on to Chaulnes to a canvas hospital, on the right side of the railway, where I was injected with anti-tetanus. Later I was again interrogated with the same result as above, except that instead of being struck, I received five days' solitary confinement in a room with no window and only a small piece of bread and a bowl of soup each day. On the fifth day I was again interrogated, and said the same as before. I said that he had no right to give me solitary confinement and that unless I were released I should, at the first opportunity, report him to the highest possible authority. The next day I was sent away and eventually reached the camp at Freiburg, where I found my brother, Captain A. E. Arnold, M.C., Tank Corps. The conduct of Gunner Ribbans and Driver Carney was beyond all praise throughout. Driver Carney drove from Villers-Bretonneux onwards.

(Signed) C. B. Arnold, Lieut.,
6th Tank Battalion,

January 1, 1919.

(This report was written by Lieut. Arnold after his return from Germany. The tank was eventually found close to the railway on the eastern side of the Harbonnières-Rosières road.)

Chapter 31
German Appreciation of British Tanks

The tardy development of both tanks and anti-tank defences has been referred to; from this it is evident that the Germans did not take kindly to the tank idea. In the tank they apparently only saw a cumbersome machine, a land Merrimac; they were unable to read the writing in iron or to understand the message that this machine brought with it on to every battlefield, namely, "the doom to all muscular warfare." Why they took so little interest in tanks may have been due to the feeling that time lacked for their development; it may have been due to the extremely low opinion held by the German Higher Command of our generalship, which prejudiced them against a purely British idea. These, however, are trivial reasons, and there must have been a deeper and broader foundation to their prejudice. The two following extracts from documents issued by the German General Staff appear to supply the real reason:

(1) From an account of the German offensive of 1918:

The use of 300 British tanks at Cambrai (1917) was a Battle of *matériel*.... The German Higher Command decided, from the very outset, not to fight a Battle of *matériel*.

(2) From an order issued by the German G.H.Q., similar to many others issued during the war:

The Higher Command is continually hearing that men who are classified as 'fit for garrison duty' are of the opinion that there is no need for them to fight and that officers hesitate to demand that they should do so. This totally erroneous assumption must be definitely and rigorously stamped out. Men in the field who are classified as fit for garrison or labour duties, but who can carry a rifle, must fight.

Such was the German tactical policy: masses of men rather than efficiency of weapons, quantity of flesh rather than quality of steel.

The policy of drafting into first-line formations men who could only just carry a rifle began in 1915. Since this date it was the constant complaint of the German regimental officer that he was obliged "to carry" in his unit an ever-increasing number of useless men—men who, for physical or moral reasons, were unfit to fight, who never intended to fight, and who never did fight.

The best men went to the machine-gun units and to the assault troops. In many cases the remainder of the infantry were of little fighting value, though many of these men might otherwise have been usefully employed in a war, which if not one of *matériel*, was at least one in which economic factors such as man-power played an important part.

By abiding by this policy of "cannon-fodder" the German Higher Command was able to look at an order of battle totalling some 250 divisions—on paper a terrific muscular force. Being pledged to a policy of employing masses of men for fighting, the Germans were not in a position to find labour for the construction of additional weapons such as tanks. It now seems clear that this policy, at least as far as tanks were concerned, was regretted before the end of the war, as the following extracts and quotations will show:

In July 1918 General Ludendorff wrote:

In all the open-warfare questions in the course of their great defensive between the Marne and the Vesle, the French were only able to obtain one initial tactical success due to surprise, namely, that of July 18, 1918. It is to the tanks that the enemy owes his success.

A similar remark was made in an order of the LIst Corps on July 23, 1918:

As soon as the tanks are destroyed the whole attack fails.

The tank victory at the Battle of Amiens brought forth a rich crop of appreciative comments from the Germans. Ludendorff on August 11 wrote:

Staff officers sent from G.H.Q. report that the reasons for the defeat of the Second Army are as follows: 'The fact that the troops were surprised by the massed attack of tanks, and lost their heads when the tanks suddenly appeared behind them, having broken through under cover of natural and artificial fog ... the fact that the artillery allotted to reserve infantry units .. was wholly insufficient to establish fresh resistance . . . against the enemy who had broken through and against his tanks.'

A 21st Infantry Divisional Order dated August 15 contained the following:

Recent fighting has shown that our infantry is capable of repelling an unsupported hostile infantry attack and is not dependent on our protective barrage.
On the other hand, a massed tank attack, as put in by the enemy during the recent fighting, requires stronger artillery defensive measures.
The duty of the infantry is to keep the enemy advancing un-

der cover of the tanks (whether infantry, cavalry, or aeroplanes) away from our artillery in order to give the latter freedom of action in its main role, *viz.*: the engagement of tanks.

This clear statement that the main duty of the artillery has become the engagement of tanks is noteworthy, especially when compared with previous orders which stated that the allotment of artillery to tank defences must not interfere with defensive barrages and counter-battery work.

This document continues:

Counter-attacks against hostile infantry supported by tanks do not offer any chances of success and demand unnecessary sacrifices; they must, therefore, only be launched if the tanks have been put out of action.

Thus, two of the mainstays of the former German defence, *i.e.* "the protective barrage" and "the immediate counter-attack," were abandoned in the event of tank attacks.

Yet one more order is interesting, that issued on August 12 to the Crown Prince's Group of Armies:

G.H.Q. reports that during the recent fighting on the fronts of the Second and Eighteenth Armies, large numbers of tanks broke through on narrow fronts, and pushing straight forward, rapidly attacked battery positions and the Headquarters of Divisions.

In many cases no defence could be made in time against the tanks, which attacked them from all sides.

Anti-tank defence must now be developed to deal with such situations.

"Messages concerning tanks will have priority over all other messages or calls whatsoever" is the last extract we will here quote, this order being sent out on September 8, 1918. These few words alone are sufficient to show that the enemy at last had awakened to the danger of the tank and was now making frenzied efforts to organise, at all costs, an efficient anti-tank defence.

It was now no longer the pluck of our Royal Air Force, the courage of our infantry, or the masses of our shells, it was the tank which threatened the German with destruction and against which he now concentrated all his energy. These efforts were, however, so belated that even the schemes and orders issued were contradictory and lacking in co-ordination; the actual practice was, needless to say, still more diverse.

From August 1918 onwards the success of almost every Allied attack was attributed to tanks in the German official communiqués. The Allies were stated to have captured such-and-such a place "by means of masses of tanks" even on occasions when very few tanks had actually been used. This explanation of any German lack of success by reference to tanks soon produced very marked results both in the German soldier and the German public.

Since the German Higher Command could explain away failure in the event of tank attack the German regimental officer very naturally came to consider that the presence of tanks was a sufficient reason for the loss of any position entrusted to his care. His men came to consider that in the presence of tanks they could not be expected to hold out. Most German officers when captured were anxious to explain that their capture was inevitable and that they had done all that could be expected of them. From this time onwards their explanations generally became very simple: "The tanks had arrived, there was nothing to be done." The failure of the Higher Command to produce tanks to combat those used by the Allies began to undermine the faith of the troops in their generals.

As a result of the "massed tank attacks," so frequently referred to in the communiques, the leading German military correspondents dealt with the tank question at considerable length. They pointed out the vital importance of tanks and inquired what the German Higher Command proposed to do about it, or reassured their readers that the situation was well in hand and that a German tank would shortly make its appearance in adequate numbers. So nervous did the press grow that the War Ministry found it necessary to offer an explanation.

General von Wrisberg, speaking for the Minister of War in the Reichstag, made the following statement:

> The attack on August 8 between the Avre and the Ancre was not unexpected by our leaders. When, nevertheless, the English succeeded in achieving a great success the reasons are to be sought in the massed employment of tanks and surprise under the protection of fog....
>
> The American Armies also should not terrify us. We shall also settle with them. More momentous for us was the question of tanks. We are adequately armed against them. Anti-tank defence is nowadays more a question of nerve than *matériel*.

On October 23 the German Wireless published the following statement by General Scheuch, Minister of War:

> Germany will never need to make peace owing to a shortage of war *matériel*. The superiority of the enemy at present is principally due to their use of tanks. We have been actively engaged for a long period in working at producing this weapon (which is recognised as important) in adequate numbers. We shall thus have an additional means for the successful continuance of the war, if we are compelled to continue it.

This statement was obviously made in reply to public criticism, but the statement that efforts were being made to produce a large number of tanks appears to be true.

It is doubtful, however, if it were true to say that they had been actively working on tanks for a long time. It is credibly reported that when Hindenburg visited the German Tank Centre near Charleroi in February 1918, he remarked, "I do not think that tanks are any use, but as these have been made, they may as well be tried." This remark of the German commander-in-chief was typical of the general feeling of the German Great General Staff towards tanks up to August 8, 1918. In our own army it also expressed precisely the feeling of a section of our Higher Command. It is hoped that, as this chapter shows the Germans were eventually, though too late, cured of their want of foresight, we have also been. As to this the future alone will enlighten us.

CHAPTER 32
Aeroplane Co-Operation with Tanks

Prior to July 1, 1918, no definite aeroplane and tank co-operation had been organised, though the want of such co-operation had been long felt, and in one of the attacks on Bourlon wood, during the Battle of Cambrai, aeroplanes had proved their value in protecting tanks from the enemy's field guns.

The assistance which aeroplanes can afford tanks falls under the two main headings of information and protection; in the future, no doubt, those of command and supply will be added.

Prior to the Battle of Arras, in February and March 1917, certain experiments were carried out in communication between tanks, aeroplanes and captive balloons by means of the Aldis daylight signalling lamp, as already mentioned in Chapter 24, these experiments did not prove a success. During the Battle of Messines aeroplanes, with considerable accuracy, reported the whereabouts of tanks on the battlefield. At the Third Battle of Ypres this useful work was continued, and again at the Battle of Cambrai, but during these last operations the days were usually so misty as to forbid much useful work being accomplished.

After the Battle of Cambrai every endeavour was made by the Tank Corps to get this co-operation regularised and placed on a sound footing, but except for some remarkable tests carried out by the 1st Tank Brigade in the vicinity of Fricourt, in February 1918, in which it was conclusively demonstrated that low-flying aeroplanes could render the greatest protective assistance to tanks, nothing was done to institute a definite system of co-operation. To do so, only one thing was required—namely, the attachment of a flight or squadron of aeroplanes to the Tank Corps for experimental purposes.

At length on July 1, 1918, some five weeks before the Battle of Amiens began, No. 8 Squadron, R.A.F., equipped with eighteen Armstrong-Whitworth machines, was attached to the Tank Corps, for the purpose of co-operating with the tanks and carrying out experiments with a view to future development. This squadron was under the command of Major T. Leigh-Mallory, D.S.O., and it was due to the energy of this officer that, in the extremely short time available, such extensive progress was made in aeroplane and tank co-operation, especially in contact work. The benefit which resulted from this co-operation cannot be over-estimated.

Early in June, No. 42 Squadron, R.A.F., had already carried out experiments with smoke flares and Very lights which were successful, whilst No. 22 Squadron attempted wireless telephony, and No. 15 Squadron visual signalling communication by means of discs swung out from the fuselage. These experiments formed the basis of the work which No. 8 Squadron now started on with the 1st, 3rd, and 5th Tank Brigades, continuing it up to the opening of the August offensive.

The account of the co-operation of No. 8 Squadron may be conveniently divided into three periods:

(1) The period of preparation, July 1 to August 8.

(2) The Battle of Amiens.

(3) The Battle of Bapaume to November 11, 1918.

During the last-mentioned period No. 8 Squadron was reinforced by No. 73 Squadron, which, being equipped with Sopwith-Camel machines, was able to deal effectively with the enemy's anti-tank guns.

The first essential of successful co-operation being comradeship, a firm alliance was at once established between the flights of No. 8 Squadron and the tank units with which this squadron was working. This was carried out by attaching tank officers to the flights, these officers frequently flying, whilst pilots and observers were given rides in the tanks.

The Battle of Hamel, on July 4, was the first occasion upon which aeroplanes were definitely detailed to work with tanks, C Flight of No. 8 Squadron being attached to the 5th Tank Brigade for this operation. The morning was a peculiarly dark one, with clouds at 1,000 ft.; nevertheless, one aeroplane managed to get off at 2.50 a.m. and a second at 3 a.m. These two machines flew low over the enemy's lines with the object of drowning the noise of the approaching tanks. Later on, another machine, flying down into the smoke of the artillery barrage, silenced some guns which were giving considerable trouble. Altogether the assistance that No. 8 Squadron rendered the Tank Corps, on this the first occasion upon which these two mechanical arms co-operated, boded well of the future.

After the Battle of Hamel, tests and training were continued, "B" Flight concentrating on wireless telegraphy and telephony with the 1st Tank Brigade, and "A" Flight on visual signalling with the 3rd Tank Brigade.

The wireless telephony tests, though of exceptional interest, did not prove very successful. Under very favourable conditions speech could be heard in a moving tank from an aeroplane flying at an altitude of 500 ft. and not more than a quarter of a mile away. It was consequently decided that, for immediate use, wireless telephony was not a practical means of communication.

Towards the end of July, a series of most successful tests were carried out in wireless telegraphy, tanks clearly receiving messages from aeroplanes at 2,500 ft. altitude and 9,000 yards away. Successful as these experiments proved, they were destined to be still-born, for time was insufficient to develop them or to apply them during active operations.

The disc signalling carried out by "A" Flight was instituted as a means of directing Whippet tanks on to their objectives. By degrees a complete code of signals was evolved so that the aeroplane was able to communicate both the nature and direction of the target. In conjunction with disc signalling, various kinds of smoke bombs and Very lights were experimented with, and by means of these several very successful manoeuvres were carried out at the Tank Gunnery School at Merlimont.

In spite of the fact that the period of preparation was too short to enable the results of tests to be applied in battle, pilots and observers had got to know a great many of the tank officers with whom they were going to co-operate, and in addition had learnt much concerning the limitations of tanks, and the kind of information required by their staffs and crews during action.

The programme of work for No. 8 Squadron on the opening day of the Battle of Amiens was as follows:

(1) Machines to fly over the line for the last hour of the tank approach march in order to drown the noise of the tank engines.

(2) Contact and counter-attack patrols to keep tank units constantly informed, by dropping messages at fixed stations, as to the progress of the battle.

(3) All machines were instructed to help the tanks whenever an opportunity arose.

On August 5 the Squadron concentrated at Vignacourt, "C" Flight being detailed to work with the 5th Tank Brigade and "B" and "A" Flights with the 4th and 3rd Brigades.

At 2.50 a.m. on August 8 three machines "took off" to cover the tanks during the last hour of the approach march. The morning was dark and the clouds appeared high. Each of these machines dropped six 25 lb. bombs, at intervals, over the enemy's lines. Between 4.50 a.m. and 5 a.m. the first four tank-contact patrol machines "took off." The valleys were already coated with thick mist and within an hour the whole country for miles was obscured. By flying very low and making use of gaps in the mist, one of these machines was able to report that tanks had passed through Demuin, and consequently it was known that the bridge there must be intact. The first message to be dropped at the Advanced Headquarters of the Tank Corps read as follows:

To Advanced H.Q. Tank Corps
    (per aeroplane).
W.4. 8th.
Machine landed 8.30 a.m. reports AAA 6.15 a.m. 4 tanks seen in action on a line 500 yards west of road through C.17.b, C.11.d, C.12.a AAA 7.15 a.m. 4 tanks seen together heading E on road beyond Hourges at C.11 central AAA 3 tanks seen together in C.6.d uncertain AAA 7.20 a.m. Green Line taken, tanks rallying to move off again AAA Foregoing report applies to 5th Tank Battalion Sector AAA 7.45 a.m. 4 tanks on road leading

north out of Demuin V.25.C.4.8 AAA 1 tank at D.1.c central AAA 4 tanks at C.11.d.3.8 heading east AAA 7.45 a.m. French infantry seen in large numbers on western outskirts of Moreuil wood and French barrage on a line C.17.c C.23.a, & C.29.a & 28.D AAA Motor transport probably armoured cars seen on road in U.26 near Domart AAA German balloon observed up just east of Caix about 8 a.m. at 1,200 ft. AAA Bombs dropped in W.22.d south of Harbonnières, target guns AAA Addressed 22nd Wing 3rd, 4th and 5th Tank Bde. Advanced Hqrs. AAA. Sent by aeroplane to dropping ground Advanced H.Q. Tank Corps.
Note added.
Cavalry and tanks in large numbers proceeding east at 8 a.m. south of Bois d'Aquenne.

Intelligence Officer,
8th Squadron, R.A.F.
8.50 a.m.

Many other such messages were dropped during the day, the Tank Brigade Headquarters being well posted with information as the attack proceeded.

On the following three days of the battle the enemy's resistance in the air became much more marked. On August 9 and 10 good targets were observed from the air in the form of large parties of infantry and transport. On the 10th, Captain West and Lieutenant Haslam were co-operating with tanks near Rosières when movement along the roads was noticed in the neighbourhood of Roye. Although some 8,000 yards from our lines Captain West immediately flew his machines in that direction and with great effect bombed and fired on the enemy's transport moving eastwards. Just as he turned to fly back, he was attacked by seven Fokker biplanes. With almost the first burst one of the hostile machines, which had got above Captain West's right-hand wing, shot his left leg off between the knee and the thigh, three explosive bullets hitting it. In spite of the fact that West's leg fell amongst the controls and that he was wounded in the right foot he managed to fly his machine back and land it in our lines. For this act of gallantry, he was awarded the Victoria Cross.

During the Battle of Amiens aeroplane co-operation had been chiefly confined to contact and counter-attack patrols. The tanks had, however, during this battle, suffered heavily from the German field guns, so, in the next great battle, the Battle of Bapaume, it was decided to make counter-gun work a feature of aeroplane co-operation.

Instead of sending all machines up on contact and counter-attack patrol, as many machines as possible were reserved for counter-gun work. From this time onwards the tendency was to concentrate more and more on this important duty, and as fresh experiences were gained this work grew more and more successful. Fortunately, just before the Third Army attack began, on August 21, No. 73 Squadron (Sopwith-Camels) was attached to the Tank Corps for this form of co-operation.

The tactics adopted in this counter-gun work are interesting. To send down zone calls was useless, as the German gunners opened fire, as a rule, when the tanks were but 1,000 yards away. Immediate action was, therefore, necessary, and this was taken by bombing and machine-gunning hostile artillery until the tanks had run over the emplacements. The method of locating the hostile gun positions consisted in carefully studying the ground prior to the attack by consulting maps and air photographs, and from this study to make out a chart of all likely gun positions.

On September 2 a most valuable document was captured which set forth the complete scheme the Germans had adopted in connection with the distribution of their guns for anti-tank work; further, in this document were described the various types of positions anti-tank gunners should take up. By the aid of this document and a large-scale map it was possible to plot out beforehand the majority of possible gun positions. As each of our aeroplanes had only about 2,000 yards of front to watch, the result was that all likely places were periodically bombed. In this way, by selecting the likely places beforehand, a great number of anti-tank guns were spotted as soon as they opened fire, and thus immense service was rendered to the tanks.

August 21 was the most disappointing day No. 8 Squadron experienced whilst attached to the Tank Corps. The morning was very foggy and it was quite impossible for the machines to leave the ground until 11 a.m., a little over six hours after zero, which was at 4.55 a.m. In spite of this the counter-gun machines were not too late to carry out useful work against several batteries; this work was chiefly carried out by No. 73 Squadron, which was quite new to the work. The value of the experience gained on this day was amply demonstrated by the effective work carried out by this Squadron on the 23rd, when many hostile guns were attacked and their crews scattered. A good example of the valuable work carried out by No. 73 Squadron occurred on September 2.

A gun was observed being man-handled towards Chaufours wood; 800 rounds were fired at it, the gun crew leaving the gun and seeking security in the wood. A little later on this crew, emerging from the wood, attempted to haul the gun into it; fire was once again opened by the aeroplane, but in spite of it the crew succeeded in their object. Bombs were then dropped on the wood, and no further movement was observed.

On September 29 a wireless-signal tank was used as a dropping station. This proved a most useful innovation, for one aeroplane dropping its message at this station found, on its return home, that this message had been received by the headquarters to which it was directed within a few minutes of it having been dropped, in fact, far quicker than it would have been had the aeroplane dropped it at the headquarters itself.

The dropping of messages to tanks in action was also successfully accomplished during the 29th. One of these messages sent down the information that the Germans were still holding the village of Bony; a group of tanks, receiving this, at once wheeled towards Bony and attacked it.

On October 8 aeroplanes once again carried out useful co-operation with the tanks. The following account is taken from the report of an aeroplane the pilot of which observed the tanks attacking Serain:

> As the tanks were approaching, we dropped bombs on various parties of Germans who were in the village. The tanks were then surrounding the village, one going right into the centre of it; a second attacked the orchard to the south, mopping up parties of Germans; whilst a third came round the north of the village and was approaching a small valley in which were 200 to 300 Germans covered by a stretch of dead ground. On seeing the tank approaching the Germans fled eastwards, whereupon we flew towards them firing our machine guns, doing great execution.

Such actions as these were of daily occurrence and they only went to prove what the headquarters of the Tank Corps had long held—namely, that the co-operation of aeroplanes with tanks is of incalculable importance, the aeroplanes protecting the tanks and the tanks protecting the infantry. In the future, no doubt, not only will messages be dropped and hostile guns silenced, but the commanders of tank battalions will be carried in the air, these officers communicating with their machines by means of wireless telephony, and supplies of petrol will be transported by means of aeroplane for the replenishment of the tanks.

CHAPTER 33
The Battle of Bapaume and the Second Battle of Arras

The operations which took place between the conclusion of the great Battle of Amiens and the signing of the armistice may conveniently be divided into three periods:

(1) The Battle of Bapaume and the Second Battle of Arras—August 21 to September 3.

(2) The Battles of Epehy and Cambrai St. Quentin—September 18 to October 10.

(3) The Battles of the Selle and Maubeuge—October 17 to November 11.

The first comprises the fighting in the devastated area, the second the breaking through of the Hindenburg system of trenches, and the third open warfare east of this system. Each of these periods will be dealt with in a separate chapter, in which most of the detail of battle preparation will be omitted so as to avoid a repetition of descriptions of work which had by now been reduced to a routine in the Tank Corps.

Towards the end of the Battle of Amiens it became apparent that the enemy was commencing to withdraw his troops in the Pusieux-Serre area opposite the Third Army front, and that, in all probability, this retirement was only part of a general withdrawal on the entire front south of the Scarpe. It was, therefore, decided on August 13 that the Third Army should prepare an attack on the German front north of the Somme, whilst the Fourth Army continued to press the enemy south of this river. Consequent on this decision the 3rd, 6th, 9th, 10th, 14th, and 15th Tank Battalions were withdrawn from the Fourth Army area, and concentrated in that of the Third. On August 15 to these battalions were added the 7th, 11th, and 12th of the 1st Tank Brigade. These moves necessitated a complete reshuffling of Tank Brigades, which on August 19 were constituted as follows:

(i) *In the Third Army Area:*

1st Tank Brigade	3rd	Battalion	Medium A.
	7th	,,	Mark IV.
	10th	,,	Mark V.
	17th	,,	Armoured cars.
2nd Tank Brigade	6th	,,	Medium A.
	12th	,,	Mark IV.
	15th	,,	Mark V Star.

3rd Tank Brigade	9th ,,	Mark V.
	11th ,,	Mark V Star.
	14th ,,	Mark V.

(ii) *In the Fourth Army Area :*

4th Tank Brigade	1st Battalion	Mark V Star.
	4th ,,	Mark V.
	5th ,,	Mark V.
5th Tank Brigade	2nd ,,	Mark V.
	8th ,,	Mark V.
	13th ,,	Mark V.

On the Third Army front the attack was to be launched on August 21 and, if successful, this attack was to be pushed forward and the front extended by an attack delivered by the Fourth Army south of the Somme. The general plan was as follows:

The VIth, IVth, and Vth Corps of the Third Army were to attack on the line Beaucourt-sur-Ancre—Moyenneville, a frontage of 17,000 yards, with the object of driving the enemy eastwards across the Arras-Bapaume road and of forcing him from the Somme area. Tanks were only to operate between Moyenneville and Bucquoy, as the ground south of this frontage was unsuited to tank movement; for this reason, no tanks were allotted to the Vth Corps.

The allotment of tanks was as follows:

VIth Corps	2nd and 3rd Tank Brigades.
IVth Corps	1st Tank Brigade

Owing to the little time available and the necessity for maintaining secrecy it was not possible to carry out any training with the divisions of the Third Army; many of these, however, had previously attended demonstrations at Bermicourt, to supplement which notes were now issued and as many lectures as possible given prior to this attack.

Another difficulty was reconnaissance, time for which was most limited. Again, previous work came to the rescue; for many officers in the Tank Corps had carefully studied the area of attack prior to the Second Battle of the Somme and had fought over it during the German spring offensive.

On the Fourth Army front the IIIrd Corps, on the left of this Army and north of the River Somme, was to attack between Bray and Albert. The 4th Tank Brigade was to assist in the attack and its machines were allotted to divisions as follows:

4th Battalion	10 tanks to 12th Division.
" "	4 tanks to 18th Division.
5th Battalion	10 tanks to 47th Division
1st Battalion	15 tanks to Fourth Army reserve

South of the Somme the 5th Tank Brigade was ordered to co-operate with the Australian Corps on the front Herleville—Chuignolles; the object of the attack being to capture these villages and the rise running east of them. Tanks were allotted as follows:

8th Battalion	12 tanks to 32nd Division
2nd Battalion	12 tanks to 1st Australian Brigade
13th Battalion	12 tanks to 2nd Australian Brigade

The Battle of Bapaume, which began on August 21, is of particular interest in that it was the first attack launched against a new tactical system of defence recently adopted by the enemy, namely, the holding of his reserves well in rear of a lightly held outpost line. In conformity with the principles of this system of defence in depth the Germans had withdrawn their guns behind the Albert-Arras railway; this eventually complicated the tank attack, for had they remained forward, as so frequently they had heretofore done, they would have been surprised and captured during the first phase of the battle; as it was, they accounted for many of our machines during the third phase.

In consequence of the enemy's new system of defence and the varying powers of the three marks of machines used by the 1st and 2nd Tank Brigades, tanks were disposed in echelons as follows:

(1) Two battalions of Mark IV tanks to operate as far as the second objective.

(2) One battalion of Mark V and one of Mark V star machines to operate against the second objective and proceed as far as the Albert-Arras railway.

(3) Two battalions of Whippets to operate beyond the Albert-Arras railway line.

Zero hour was at 4.55 a.m. The Mark IV battalion moved forward and successfully cleared the first objective and pushed on towards the second. Once again, the attack was a surprise, perhaps not so much through it being unexpected as through the inability of the Germans to meet it, especially as their guns had been withdrawn. To illustrate how complete this surprise was it is only necessary to mention that candles were found still burning in the trenches when we crossed them, and papers and equipment scattered broadcast gave evidence of the hurried flight of the enemy. The second echelon and the Whippets had, however, a far more difficult task to accomplish.

The Albert-Arras railway had, previous to the attack, been turned into a strongly defended line of machine-gun nests covered by the

German guns east of it. Unfortunately, the thick ground mist, which had shielded the approach of the first echelon, now began to lift; this enabled the German artillery observers to direct a deadly fire on the tanks, in fact, each individual tank became the centre of a zone of bullets and bursting shells. Avoiding these zones, our infantry pushed on with few casualties. During this day's fighting many parties of the enemy, some over a hundred strong, surrendered *en bloc* directly the tanks were seen approaching. Such action on the part of the German infantry was becoming a stereotyped procedure in all tank attacks. On the 21st, of the 190 tanks which took part in the attack, 37 received direct hits.

On August 22 the IIIrd Corps launched its attack on a front of some 10,000 yards with complete success. The tanks, which had been instructed by the IIIrd Corps to proceed in rear of the infantry, in actual fact led the attack the whole way, effecting a penetration of about 4,000 yards. All objectives were gained, and at the end of the day our line ran east of Albert, east of Meaulte, east of the Happy Valley, and through the western outskirts of Bray-sur-Somme.

On the following day the attack of the IIIrd Corps was continued in conjunction with that of the Third Army to the north and the Australian Corps to the south. The IIIrd Corps captured Tara and Usna hills, employing six tanks of the 1st Tank Battalion in this action. On the Australian front the thirty-six machines of the 5th Tank Brigade deployed and led the infantry right on to their objective, which was successfully occupied. On reaching this the machines of the 2nd and 13th Battalions exploited north of Chuignolles with the 3rd Australian Brigade, whilst those of the 8th Battalion rallied. During this attack the enemy put up a stout resistance, his machine-gunners fighting with great spirit and in many cases continuing to fire their guns until run over by the tanks. Curiously enough, in comparison with this, on the previous day the enemy's machine-gunners on the IIIrd Corps front scarcely put up any fight at all, and when asked why they had not done so replied: "Oh! it would not have been any good."

The following is a typical battle-history sheet depicting the tank fighting at this period; it was written by a tank commander who took part in the above attack.

> At 4.25 a.m. on the 23rd instant, I proceeded with my female tank 'Mabel' (No. 9382) in front of the infantry. I made a very zigzag course to the wood in the south-west edge of the village

of Chuignolles, where I encountered an anti-tank gun which was eventually knocked out by the male tank commanded by 2nd Lieut. Simmonds, who was operating on my right. I then worked up the south side of the village, heavily machine-gunning all the crops and copses, and dislodged several machine-gun crews of the enemy. I next came back to the village and mopped up the enemy on the outskirts until it was clear. Then, emerging from the smoke of two shells which dropped short, I found myself in the midst of a battery of whizz-bangs. The gunners of this battery I at once proceeded to obliterate with good success, after which I came behind another battery and proceeded with the same operation.

Then I started to take infantry over to the high ground south of Square wood in L.35d, when I was called back by an Australian colonel to attend to some M.G. nests which had been left in the centre of the village; here I mopped up twelve M.G. nests and then started to catch up the remainder of the tanks and the barrage, but while at the top of a very steep bank, approximately R.10.b.30.50, I received a direct hit from a whizz-bang on the front horns, which sent me out of control to the bottom of the bank, where I found it had broken one plate of my left track. After repairing this, the barrage having finished, and as tanks were coming back to rally, I brought my tank back to the rallying-point at Amy wood.

In the Third Army the attack was re-opened on August 23 by a moonlight operation, starting at 4 a.m., carried out against the village of Gomiecourt. The 3rd Division, supported by ten Mark IV tanks of the 12th Battalion, attacked this village and carried it. A little later the Guards Division with four Mark IV.s captured the village of Hamelincourt.

At 11 a.m. the VIth Corps, assisted by fifteen Whippets, attacked in the direction of Ervillers—Behagnies—Sapignies, and by noon were east of the Bapaume-Arras road. Near Sapignies heavy machine-gun fire was encountered, which prevented our infantry moving forward; notwithstanding this the Whippets continued their advance. In one machine the officer and sergeant were killed; the remaining man, however, after placing the corpses in a shell-hole, continued single-handed to follow the other tanks, and when a target offered itself, locked his back axle and fired his Hotchkiss gun. Although Sapignies and Behagnies were not captured the operation was successful in securing a large number of prisoners. This attack materially assisted that of the IVth Corps on Achiet-le-Grand and Bihucourt, which were captured by the 37th Division and six Mark V tanks of the 1st Tank Brigade.

At 5.7 a.m., in spite of a very heavy gas barrage which took place between 2 a.m. and 4 a.m., eighteen tanks of the 11th Battalion and eight of the 9th Battalion co-operated with the 52nd and 56th Divisions in an attack on the Hamelincourt-Heninel spur. Both these objectives were carried with small loss.

On August 24 the attacks of the Third and Fourth Armies were pushed with vigour, the 1st, 3rd, and 4th Tank Brigades co-operating.

At 2 a.m. on the IVth Corps front four Mark IV machines assisted the 37th Division west of Sapignies; later on, seven Mark IV tanks and nineteen Whippets attacked with the New Zealand and 37th Divisions on Grevillers, Biefvillers, and Loupart wood, which were captured. On the front of the 56th Division a bloody engagement began at 7 a.m. for the ownership of the line St. Leger—Henin-sur-Cojeul, eleven tanks of the 11th Battalion co-operating after an approach march of 10,000 yards. All objectives were gained. At 2 p.m. this attack was continued, two tanks leading the infantry north of Croisilles as far as the Hindenburg Line, which was strongly held.

Both these machines had particularly exciting experiences. One received a direct hit which rendered the officer commanding it unconscious; recovering his senses, he at once took charge of his machine and pushed on. In the other the crew were forced to evacuate the tank on account of the enemy surrounding it with phosphorus bombs; before leaving it the officer in command turned the tank's head towards home and then, getting out, walked between the front horns of the machine until the fumes had cleared away. All this time the tank was surrounded by the enemy. This attack failed; it is interesting, however, to note that one tank of the 11th Battalion covered during this operation 40,000 yards in some twenty-six hours.

At 3.30 p.m. five tanks of the 9th Battalion took part in an attack on the line Mory copse—Camouflage copse. At first there was not much opposition, but after Mory copse was reached the enemy put up a stout resistance and, refusing to surrender, was killed almost to a man, one party of about sixty being put out of action by four rounds of 6-pounder case shot.

At dawn this day on the Fourth Army front five tanks of the 1st Battalion assisted the 47th Division in the recapture of the Happy Valley, which had been lost on the previous afternoon. This attack was entirely successful, and besides the Happy Valley the extensive village of Bray was added to our gains.

On the evening of August 24th, the 3rd Tank Brigade was transferred from the Third Army to the Canadian Corps of the First Army for the forthcoming attack on the First Army front, which was to initiate the Second Battle of Arras. This entailed a lengthy night march of 29,000 yards from Blaireville and Boisleux-au-Mont to Moyenneville, thence to Achicourt.

On August 25 minor tank attacks were carried out on the IVth Corps front against Favreuil, Avesnes, Thilloy, and Sapignies, and on the 26th tanks of the 9th Battalion, some of which had moved 37,000 yards since the night of the 24th–25th, attacked with the Guards Division north of Mory. This attack was not a great success, due to the dense mist, which made co-operation almost impossible and the maintenance of direction most difficult. During this engagement one tank had five members of its crew wounded by anti-tank rifle bullets.

On the Canadian Corps' front an attack was carried out opposite Fampoux and Neuville-Vitasse, the 2nd and 3rd Canadian Divisions operating with tanks of the 9th and 11th Divisions against Wancourt, Guemappe, and Monchy-le-Preux. Near Monchy several tanks were knocked out, the crews joining the infantry to repel a local counter-attack. The sergeant of one crew, hearing that the enemy had captured his tank, collected his men and charged forward to recover it, arriving at one sponson door of the machine as the enemy were scrambling out of the opposite one.

On the following day, the 27th, operations continued east of Monchy-le-Preux, and in the Guemappe-Cherisy area, but ceased altogether as far as tanks were concerned until the 29th. On this day minor tank operations were carried out on the Third Army front south-west of Beugnatre by the 1st Brigade, the enemy having evacuated Thilloy and Bapaume. This Brigade, on the 30th, co-operated with the 5th British and New Zealand Divisions against Fremicourt, Beugny, Bancourt, Haplincourt, and Velu wood, whilst the 2nd Tank Brigade attacked Vaux-Vraucourt. All these attacks were successful.

On the last day of August 1918, probably the most decisive month in the whole war, nine Mark IV.s of the 12th Battalion and four Whippets of the 6th Battalion attacked the Longatte trench, Moreuil switch, and Vraucourt trench, taking all these objectives, and on the following day, September 1, Whippets of the 6th Battalion completed the above operations by establishing the infantry on the slopes east of Vaux-Vraucourt.

The Second Battle of Arras reached its zenith on September 2 when the famous Drocourt-Queant line, which we had failed to reach in April 1917, was broken. Starting from the south the 1st Tank Brigade operated with the 42nd and 5th Divisions against Beugny and Villars-au-Flos. To the north of this attack the 2nd Tank Brigade assisted in the VIth Corps operations against Moreuil, Lagnicourt, and Morchies. This attack was made in conjunction with those of the Canadian and XVIIth Corps against the Drocourt-Queant line. This line was attacked by the 1st and 4th Canadian Divisions and the 4th Division, together with as many tanks as the 9th, 11th, and 14th Battalions of the 3rd Tank Brigade could muster. The assembly of these machines was difficult owing not only to the intricate nature of the Sensée valley but to the fact that active operations were taking place throughout these preparations.

The Drocourt-Queant line, built in the spring of 1917, was protected by immensely strong belts of wire entanglement, and it was expected that every effort would be made on the part of the enemy to hold these defences at all cost; nevertheless, on the whole, less opposition was encountered than had been anticipated. Except for anti-tank rifle fire, which was especially noticeable at Villers-les-Cagnicourt, the tanks met little opposition. It is estimated that in this attack one company of tanks alone destroyed over seventy hostile machine guns, the German gunners surrendering to the tanks as they approached.

On the next day, the enemy falling back, the Whippet tanks pushed forward to Hermies and Dermicourt. Thus, the Second Battle of Arras ended in an overwhelming success by the piercing of the renowned Drocourt-Queant line. A blow had now been delivered from which the enemy's moral never recovered.

Since August 21, in all, 511 tanks had been in action, and except for one or two minor failures every attack had culminated in a cheap success—cheap as regards our own infantry casualties, especially so when it is remembered that during the fortnight which comprised the Battle of Bapaume and the Second Battle of Arras no fewer than 470 guns and 53,000 prisoners were captured. Thus, in a little less than one month the German Army had lost to the First, Third, and Fourth British Armies 870 guns and 75,000 men without counting killed and wounded.

Chapter 34
German Anti-Tank Tactics

From September 1916 onwards to the conclusion of the war, German anti-tank tactics passed through three phases. Firstly, the enemy had no anti-tank defence at all, or what he devised he based upon a misconception of what the tank could accomplish. Secondly, having learnt but little about tanks, he considered that only a small expenditure of effort and *matériel* was required to deal with weapons of so limited a scope. Thirdly, from August 1918 onwards, he took panic and over-estimated their powers; his efforts at anti-tank defence became feverish and he appeared to be willing to make any and every sacrifice to combat this terrible weapon.

Captured documents clearly show that the introduction of tanks was as great a surprise to the German General Staff as to their fighting troops. It is true that certain vague rumours had been circulated that the Allies might use some new weapon, but, as such rumours have throughout the war been current on all sides, no particular importance was attached to them. In spite of the fact that tanks were used on several occasions between September 15 and November 13, 1916, and that the enemy held in his possession near Gueudecourt a captured tank for some fourteen days, he formed a most inaccurate idea of it.

During the winter of 1916 and 1917 instructions were issued on anti-tank defence. These were based on the following entirely erroneous ideas:

(1) That tanks were largely dependent on roads.

(2) That tanks would approach the German lines in daylight.

(3) That tanks were impervious to machine-gun fire.

These led to the Germans depending on road obstacles such as pits and indirect artillery fire; as a matter of fact, at this time the most potent weapon which could have been used against tanks was the machine gun firing A.P. bullets. That the Mark I tank was not proof against these bullets was not discovered until April 1917, after the British failure at Bullecourt. This discovery was of little use, for by the time the next battle was fought, Messines, a tank with thicker armour, the Mark IV, had replaced the Mark I.

It is evident that throughout this period the German Higher Command gave little thought to the tank question and quite failed to appreciate the possibilities of the machine. Prisoners were questioned and rough sketches, many grotesque in the extreme, were obtained from them and published for information. What information they imparted was misleading; in fact, the whole attitude of the German General Staff, during this period, may be summed up as "stupidity tempered with ridicule."

During 1917 the German grew to realise that artillery formed the chief defence against tanks. Great prominence was given to indirect fire by all types of guns and howitzers, and in spite of several dawn attacks the enemy laid great stress on what he called "Distant Defence." As actual operations proved, indirect artillery fire produced little effect save on broken-down machines. Partially learning this, the Germans resorted to special anti-tank guns, and on an average, two, protected by concrete, were emplaced on each divisional front; these were in certain sectors supplemented by captured Belgian and old German guns. Fixed anti-tank guns proved, however, of little use, for though a few tanks were knocked out by them, notably at Glencorse wood on the first day of the Third Battle of Ypres, they generally were destroyed by our terrific initial bombardments. Curiously enough, though both indirect fire and fixed guns proved a failure, little consideration was, at this time, given to the simplest form of artillery action against tanks, namely—direct fire by field guns.

Infantry anti-tank defence, during 1917, was negligible and chiefly consisted in instructions how to "keep its head" and to leave the rest to the artillery; the use of bundles of stick bombs was recommended, and though A.P. bullets received no great support, the effect of the splash of ordinary S.A.A. was not realised in the least degree.

Prior to the Battle of Cambrai (November 1917) the true anti-tank defence had been mud, mud produced by gunfire and rain. At this battle the enemy was caught completely unawares, his anti-tank defence was slight; but a feature of the operation was the improvised defence put up by a few of the enemy's field guns, which inflicted heavy casualties on tanks, especially at Flesquières. In general, however, the enemy realised the ineffectiveness of his anti-tank defence, yet curiously enough he at present showed no decided inclination to adopt direct field-gun fire as its backbone.

In spite of the fact that the incident of a German battery, served by an officer, putting a considerable number of tanks out of action had received mention in the British dispatches, still the enemy remained oblivious to the utility of direct fire, and in place of praising his gunners, as we did, he praised troops from Posen who had put up a determined resistance against tanks in Fontaine Notre Dame.

The German counter-attack on November 30, 1917, which resulted in the capture of a considerable number of tanks, seems to have entirely allayed any anxiety created by the attack on the 20th; for though the question of anti-tank defence was given rather more prominence than heretofore, no greater practical attention was paid to it during the winter of 1917–1918 than during the one previous to it.

The German offensive in the spring of 1918 put all defensive questions into the background. This period, however, produced a new weapon, the German anti-tank rifle.

This rifle was first captured during the Battle of Hamel on July 4. It had only just been issued to certain divisions; other divisions were equipped with it later on.

This weapon was 5 ft. 6 in. in length, it weighed 36 lb. and fired single shots, using A.P. ammunition of ·530 calibre. It was too conspicuous and too slow a weapon to be really effective against tanks, though it could easily penetrate them at several hundred yards range. Its chief disadvantage was that the German soldier would not use it; not only was he not trained to do so, but he was afraid of its kick, and still more afraid of the tanks themselves. It is doubtful if 1 *per cent.* of the anti-tank rifles captured in our tank attacks had ever been fired at all.

The French counter-attack between the Aisne and the Marne in July, followed by the British victories of Amiens and Bapaume in August, struck through the opacity of the German General Staff like a bolt from out the blue, with a result that a complete *volte face* was made as regards tanks. The instructions now issued gave anti-tank defence the first place in every project; the eyes of General Ludendorff were now opened, and, realising the seriousness of the tank problem, on July 22 he wrote as follows:

> The utmost attention must be paid to combat tanks—our earlier successes against tanks lead to a certain contempt for this weapon of warfare. We must, however, now reckon with more dangerous tanks.

This is a more human document than those subsequently issued by the German Chief of the General Staff. Ludendorff now clearly realised that anti-tank defence had been neglected; he probably realised also that this neglect would be difficult to explain to the army and the public, which, as a result of failures, were about to become far more critical of their leaders than ever before.

It is not clear, however, whether Ludendorff realised a still more serious aspect of the tank problem, namely, that it was now too late to organise an efficient defence against the "more dangerous tanks." Such a defence might have been created before these tanks were available in effective numbers; it could not be organised now unless the pressure the Allies were now exerting could be relieved. This was impossible, for the motive force of this pressure was the tank.

The steps which the German General Staff now took to combat the tank are interesting. Special officers were appointed to the staffs of Groups of Armies, Corps, Divisions, and Brigades, whose sole duty it was to deal with anti-tank defence within these formations. The field gun was at length recognised as the most efficient anti-tank weapon available. These guns were organised as follows:

(1) A few forward and silent guns in each divisional sector—outpost guns.

(2) Sections from batteries in reserve were allotted definite sectors. On a tank attack taking place, they would gallop forward and engage any tank entering the sector allotted to the section. These sections of guns proved the backbone of the German anti-tank defence.

(3) All batteries (howitzers included) were ordered to take up positions from which advancing tanks could be engaged by *direct fire*. The most effective range for this purpose was first considered to be over 1,000 yards; this was gradually reduced to about 500.

Batteries in (1) and (2) were to be employed for anti-tank work only, in (3) they were available for other work, but in the event of a tank attack the engagement of tanks was their chief task.

The duty still allotted to the infantry was "to keep their heads" or "to keep calm," actions which at this period were impossible to the German Higher Command directly tanks were mentioned. Other orders laid down that in the event of a tank attack "infantry should move to a flank." How this was to be done when tanks were attacking on frontages of twenty to thirty miles was not explained. A.P. ammunition had to a great extent fallen into discredit, and, curious to record, the effect of "splash" as a means of blinding a tank was still hardly realised, and this after two years of tank warfare.

As artificial obstacles had proved of little use from the end of July, when the Germans withdrew behind the Rivers Ancre and Avre, until the signing of the armistice every effort was made to use river lines as a defence against tanks. Road obstacles and stockades were still in use, but though they proved a hindrance to the movement of armoured cars they proved none to tanks.

A great deal of energy and explosive material was expended in laying minefields. At first, special mines in the form of a shallow box were used; later on these were replaced by shells. Lack of time, however, prevented the enemy from developing sufficiently large minefields to produce an important result.

The idea of combining the various forms of anti-tank defence under one command in such a way as to form an anti-tank fort had been dealt with on paper, but was only in a very few cases put into practice. The idea was a sound one, and if well combined with natural obstacles it would have formed the best defence against tanks that the enemy could have created with the means at his disposal.

An anti-tank fort was to consist of:

Four field guns, 2 flat-trajectory *minenwerfer*, 4 anti-tank rifles, and 2 machine guns firing A.P. ammunition. The fort was to be sited several thousand yards behind the outpost guns and close to the main line of defence.

Throughout the last two years of the war occasional successes were gained by the Germans by various means of anti-tank defence, these usually being due to a combination of the following circumstances:

(1) The use of tanks outside their limitation.

(2) A hitch or failure in carrying out the plan of attack.

(3) An exceptional display of resource, initiative, and courage on the part of some individual German soldier.

In general, the keynote of the German anti-tank defence was lack of foresight, the development of tanks not being appreciated. Among the very large number of captured orders dealing with anti-tank defence there is no recorded instance of any anticipation of superior types of tanks to those already in use. The German General Staff lacked imagination and the faculty of appreciating the value of weapons that had not been explained to them whilst at school; obsolescence dimmed their foresight.

Chapter 35

The Battles of Epehy and Cambrai—St. Quentin

On September 4 all Tank Brigades were withdrawn from Armies and placed in G.H.Q. reserve to refit and reorganise. When this had been completed Tank Brigades were constituted as follows:

1st Tank Brigade	7th Battalion	Mark IV.
	11th ,,	Mark V Star.
	12th ,,	Mark IV.
	15th ,,	Mark V Star.
2nd Tank Brigade	10th ,,	Mark V.
	14th ,,	Mark V.
3rd Tank Brigade	3rd ,,	Medium A.
	6th ,,	Medium A.
	9th ,,	Mark V.
	17th ,,	Armoured cars.
4th Tank Brigade	1st ,,	Mark V.
	4th ,,	Mark V.
	5th ,,	Mark V.
	301st American Battalion	Mark V Star.
5th Tank Brigade	2nd Battalion	Mark V.
	8th ,,	Mark V.
	13th ,,	Mark V.

At 7 a.m. on September 17, in a heavy storm of rain, the Fourth and Third Armies initiated the Battle of Epehy by attacking on a front of some seventeen miles from Holnon to Gouzeaucourt, the First French Army co-operating south of Holnon.

On September 18 the 4th and 5th Tank Brigades were released from G.H.Q. reserve and allotted to the Fourth Army, the 2nd Tank Battalion having been transferred to this Army on September 13.

On this day the Battle of Epehy continued on the front Epehy-Villeret, some 7,000 yards long. In this attack twenty tanks of the 2nd Battalion assisted the IIIrd Corps, Australian Corps, and IXth Corps. On the IIIrd Corps front heavy machine-gun fire was encountered and overcome, many machine guns being destroyed. On that of the IXth progress was slow, and the Australians, meeting with little resistance, captured Ronssoy and Hargicourt.

After two days' rest the attack was continued on the 21st, nine tanks of the 2nd Battalion operating on the IIIrd Corps front against the Knoll and Guillemont farm. Two of these machines carried forward infantry, but the machine-gun fire was so heavy that it was not possible to drop them. During this day the enemy put up a most determined resistance and there were not sufficient tanks engaged to silence his machine guns. Another two days' rest followed, and then again was the attack renewed on the IXth Corps front against Fresnoy-le-Petit and the Quadrilateral, nineteen machines of the 13th Battalion attacking with the 1st and 6th Divisions.

So heavy was the enemy's gas barrage on this day that some of the tank crews were forced to wear their respirators for over two hours on end. In spite of the enemy being in great strength eighteen machines assisted the infantry. Thus, ended the Battle of Epehy and though the advance was not great nearly 12,000 prisoners and 100 guns were added to the "bag."

Preparations were now set in hand for an extensive attack against the Hindenburg and auxiliary lines of defence, which together formed a zone of entrenchments for the most part very heavily wired and extending over a depth varying from 8,000 to 16,000 yards. This attack entailed another hasty reorganisation of tank battalions, which was completed by September 26, when the battle order of brigades was as follows:

1st Tank Brigade	7th Battalion	Mark IV	Bullecourt.
H.Q. Bihucourt	11th ,,	Mark V Star	Barastre.
	12th ,,	Mark IV	W. of Ruyaulcourt.
	1st T.S. Coy.		S. of Velu.
	2nd G.C. Coy.		Bancourt.
2nd Tank Brigade	10th Battalion	Mark V Star	Auchy les Hesdin.
H.Q. Gomiecourt	14th ,,	Mark V	Winnipeg Camp.
	15th ,,	Mark V Star	Hermies.
	2nd T.S. Coy.		Gomiecourt.
	1st G.C. Coy.		N.W. of Vaux-Vraucourt.
3rd Tank Brigade	5th Battalion	Mark V	E. of Cartigny.
H.Q. Barleux	6th ,,	Medium A	S. of Tincourt.
	9th ,,	Mark V	S. of Tincourt.
	3rd T.S. Coy.		S. of Tincourt.
4th Tank Brigade	1st Battalion	Mark V	Manancourt.
H.Q. Templeux-la-Fosse	4th ,,	Mark V	S. of Manancourt.
	301st American Battalion	Mark V Star	S. of Manancourt.
	4th T.S. Coy.		S. of Manancourt.
5th Tank Brigade	2nd Battalion	Mark V	Suzanne.
H.Q. Bois-de-Buire	3rd ,,	Medium A	S. of Roisel.
	8th ,,	Mark V	S. of Tincourt.
	13th ,,	Mark V	S. of Tincourt.
	16th ,,	Mark V Star	S. of Tincourt.
	17th ,,	Armoured Cars	Buire.
	5th T.S. Coy.		S. of Tincourt.

The rapidity with which these changes were made would, a few months back, have bewildered both the Tank Corps Headquarter Staff and the brigade and battalion commanders themselves; now the knack of rapid movement had been mastered, and though great energy had to be exerted during such reorganisations, they were generally accomplished in time and efficiently.

On September 27 the great battle began, comprising the First,

Third, and Fourth Armies on a front of sixteen miles. The battlefield was divided into two main sectors, to the north that of the First and Third Armies, between the Sensée River and Gouzeaucourt, with the object of capturing Bourlon Hill, and to the south that of the Fourth Army with the capture of the Knoll, Guillemont farm, and Quennemont farm as its objectives.

East of the First Army front line ran the canal Du Nord, a formidable obstacle to tanks in spite of the fact that it was dry, having never been completed. This canal varied from 36 to 50 ft. wide at the bottom, and was 12 ft. deep, and the slopes of its sides were in many places steep. The enemy, evidently suspecting that tanks might attempt to cross it, had at certain places rendered this temporarily impossible, so it was thought, by cutting in its bank a vertical wall 9 ft. deep for several hundreds of yards along the eastern side between Mœuvres and Inchy. In the Maquion-Bourlon sector the enemy had made little anti-tank preparation, probably considering that the canal itself formed a sufficient obstacle. In the Beaucamp sector, however, anti-tank preparations were exceptionally thorough, many anti-tank rifles being placed in position here.

Sixteen tanks of the 7th Battalion, all Mark IV.s, some of which had fought over very nearly the same ground in November 1917, were allotted to co-operate with the Canadian Corps. In spite of the formidable great ditch which lay in front of them, fifteen of these machines crossed the canal Du Nord near Mœuvres, and attacked Bourlon village and the western edge of Bourlon wood. Of these fifteen machines only three were put out of action, one by a mine placed in a road leading through a gap in the canal and two by a battery near Deligny Hill.

On the Third Army front Corps were disposed from north to south as follows: XVIIth, VIth, IVth, and Vth Corps.

Twenty-six tanks of the 15th Battalion operated with the XVIIth Corps south of Bourlon wood, and with the VIth Corps against Flesquières and Premy Chapel. A fine performance was here carried out in crossing the canal, and although more than one attempt had to be made by several of the tanks the 9 ft. wall was successfully surmounted. This attack was an overwhelming success in spite of the heavy tank casualties, 11 out of the 26 machines operating being hit on the extreme objectives. On the IVth Corps front 12 machines of the 11th Battalion attacked between Gouzeaucourt and Trescault: this operation was, however, only partially successful.

On the front of the Fourth Army the 27th American Division, supported by twelve tanks of the 4th Tank Battalion, carried out a preparatory attack on the Knoll, Guillemont and Quennemont farms, the object being to advance the front line so as to be in a better position to attack in force on the 29th. The Germans holding this sector of their line were reliable and well led troops, and in spite of the fact that the tanks and infantry reached their objectives a counter-attack drove them back, with the result that up to zero-hour on the 29th the actual location of our front line was very uncertain.

On September 28 a small local attack, which was completely successful, was carried out against Raillencourt and St. Olle; in this, six tanks of the 7th Battalion took part.

On the following day seven tanks of the 11th Battalion co-operated with the Vth Corps in the capture of Gonnelieu and Villers Guislain in spite of strong resistance put up by the enemy.

On the Fourth Army front an important Battle of considerable magnitude was fought on the 29th, involving some 175 tanks. The object of this battle was to force the Hindenburg Line between Bellenglise and Vendhuile. Along this front is situated the St. Quentin canal, and as, between Bellicourt and Vendhuile, this canal runs underground through a tunnel, it provided the German garrisons of this sector of their line with good underground cover. An operation on this sector had been the subject of careful study by the Tank Corps General Staff both in England and France ever since the summer of 1917, as this tunnel and a shorter one just north of St. Quentin provide the only negotiable approaches for tanks over the canal. It was fully realised that the enemy would put up a most determined resistance to secure his retention of the tunnel, for, should it be occupied by us the whole of the Hindenburg defences north and south of it would be threatened.

The attack was to be carried out by four corps, the IXth Corps on the right, the American and Australian Corps in the centre, and the IIIrd Corps on the left.

The American Corps was to capture the first objective, the strongly entrenched system east of Bony; the Australian Corps was then to pass through the gap made and be followed by the American Corps exploiting north and south. The IXth Corps was to clear the west bank of the St. Quentin canal under cover of the southern wing of the American exploitation force, whilst the IIIrd Corps was to move forward with the left of the American Corps.

Tanks were allotted to Corps as follows:

3rd Tank Brigade, 5th, 6th, and 9th Battalions, to the IXth Corps.

4th Tank Brigade, 1st, 4th, and 301st American Battalion, to the Australian Corps.

5th Tank Brigade, 8th, 13th, and 16th Battalions, in Army reserve.

The 301st American Battalion was attached for operations to the 27th American Division.

A thick mist covered the ground when the tanks moved forward at 5.50 a.m. It will be remembered that the situation opposite the Knoll and the two farms of Guillemont and Quennemont was very obscure. This attack, which was to break the well sited and highly organised Hindenburg Line, was necessarily a "set piece" attack in which objectives, allotment of tanks, etc., had to be carefully worked out beforehand. The plan of operations was based on the assumption that the line—the Knoll—Guillemont farm—Quennemont farm—would form the "jumping off" line. The resistance put up by the enemy in this sector was far greater than ordinary, with the result that up to the time of the attack the above line was still in German hands.

This meant that the artillery programme would have to be hastily changed or left as it was. The latter course was decided on so as to obviate confusion, and this necessitated the infantry attackers starting at a considerable distance in rear of the protective barrage. As events turned out the task set the Americans proved too severe, nevertheless with great gallantry they pushed forward, some of them actually forcing their way through the German defences. The majority, however, were mown down by the exceptionally heavy machine-gun fire which was brought to bear on them. The attack failed.

Meanwhile the 301st American Tank Battalion met with a disaster, for, whilst moving forward from near Ronssoy, it ran into an old British minefield west of Guillemont farm laid in the previous February; ten machines were blown up and only two succeeded in assisting the infantry. This minefield consisted of rows of buried 2 in. trench-mortar bombs, each containing 50 lb. of ammonal; the explosions were terrific, the whole bottom of many machines being torn out; in nearly all cases the crews of these tanks suffered very heavy casualties.

In the south, tanks of the 4th and 5th Brigades cleared Nauroy and Bellicourt and broke through the Hindenburg Line. The mist now began to lift, and consequent on the failure of the northern attack, the attackers were placed, tactically, in a very dangerous situation, for the enemy was now able to fire into their backs. Several tanks, which had been allotted to later objectives, on realising the seriousness of the situation went into action on their own initiative without either artillery or infantry support. This very gallant action undoubtedly saved a great many infantry casualties, though the tanks themselves suffered heavy losses.

On the right the attack of the IXth Corps was a complete success; in the first rush the 46th Division crossed the canal, a magnificent performance, and captured Magny and Etricourt with 4,000 prisoners. The tanks operating with this corps, being unable to cross with the troops, who waded and swam the water in the canal, moved on Bellicourt, a difficult operation in the dense fog. From this place they swung south, working down the bank of the canal, and arrived in time to take part in the capture of Magny. During this action the enemy's artillery fire proved very accurate; which was, however, to be expected, for this was the third tank assault on the Knoll—Guillemont—Quennemont position; consequently, the German gunners had become thoroughly drilled in the defence of this sector.

On the following day eighteen tanks of the 13th Battalion worked up the Hindenburg and the Nauroy—Le Catelet lines, but on account of some misunderstanding the infantry did not follow, consequently the operation did not prove of much value.

On the First Army front six tanks of the 9th Battalion operated with the 3rd and 4th Canadian Divisions against Cuvillers, Blecourt, and Tilloy; they crossed the Douai-Cambrai road near Sancourt and greatly helped the infantry by overcoming the determined machine-gun resistance which was encountered throughout this attack. On the next day further tanks of this Battalion assisted the 32nd Division in occupying the Fonsomme line east of Joncourt. In this action smoke clouds were used from tanks to cover their approach from the observation of the German gunners; this proved very successful and undoubtedly reduced loss by gun fire. One tank had a curious experience: a smoke bomb having burst on the top of it, the crew were forced to evacuate the machine on account of the fumes being drawn inside.

The tank commander, having put the fire out, was unable to find his crew; as time was short, he got inside the tank and continued his advance alone; on his way forward, he took on board an officer and two men of the Manchester Regiment. The tank then went into action against a machine-gun nest; as the improvised crew was ignorant of the Hotchkiss gun each time a jam occurred the tank commander had to leave the driver's seat to rectify it. Shortly afterwards the truant crew turned up, so the tank commander, having first driven his newly-made comrades to cover, dropped them, and then proceeded on his way.

On October 3 an attack was launched against the Sequehart-Bony front in which twenty machines of the 5th Tank Brigade proved of very great assistance to the 32nd and 46th Divisions. Sequehart was cleared and so was Ramicourt and Doon copse, but Montbrehain remained uncaptured.

On the 4th, the 3rd Tank Battalion was transferred from the 5th to the 3rd Tank Brigade, and a day later the 16th Tank Battalion from the 5th to the 4th Tank Brigade. The 8th, 9th, and 18th Battalions were withdrawn into G.H.Q. reserve. On the 5th the first phase of the Battle of Cambrai—St. Quentin opened with a failure to take Beaurevoir in which attack six tanks of the 4th Tank Battalion attempted to assist the 25th Division. Co-operation in this action was indifferent, due chiefly to the fact that the infantry of this division had never been trained to work with tanks. This failure was partially retrieved by a brilliantly executed attack by the Australians supported by twelve tanks of the 16th Battalion against Montbrehain. This village was held by the enemy in strength, and many good targets at close quarters were obtained for 6-pounders firing case shot. The co-operation throughout was excellent, as, since the Battle of Hamel, had always been the case when operating with the Australian Corps—tank commanders constantly getting out of their tanks and talking to the infantry.

The second phase of the Cambrai—St. Quentin battle opened on the morning of October 8 on an eighteen miles front—it was entirely successful. Tanks were allotted as follows, eighty-two in all being used:

1st Tank Brigade	12th Battalion	1 Company to IVth Corps.
		1 Company to VIth Corps.
		1 Company to XVIIth Corps.
	11th Battalion	To Vth Corps.
4th Tank Brigade	1st Battalion	To XIIIth Corps.
	301st American Battalion	To IInd American Corps.
3rd Tank Brigade	3rd Battalion	To IInd American Corps.
	6th Battalion	To IXth Corps.

The attacks, carried out by the 12th Tank Battalion on the front Niergnies—La Targette, were successful, the infantry universally testifying to the assistance rendered by this battalion. An interesting encounter now took place, the enemy counter-attacking from the direction of Awoingt with four captured British Mark IV tanks, one male and three females. The counter-attack was speedily dealt with, the renegade male being knocked out by a 6-pounder shell fired by one of our own tanks and one female put out of action by a shell fired from a captured German field gun by a tank section commander; the remaining two females fled on the approach of one of our machines of the same sex. So ended the second tank encounter as successfully as the first, which it will be remembered was fought near the village of Cachy on April 24, 1918.

The other actions fought on this day were briefly as follows:

One Company of the 11th Battalion assisted the 32nd Division against Villers-Outreaux, another company operated with the 21st Division and the third company with the 38th. This last company was of great assistance, as the infantry had been held up by a broad belt of wire which they were unable to cross until the tanks crushed down pathways through it.

The 6th Tank Battalion, operating with the IInd American Corps, carried out its programme, one of its machines putting three batteries of field guns out of action in Fraicourt wood; and the 3rd Battalion came into action in the neighbourhood of Serain. This village was very strongly defended, the enemy holding it to cover his withdrawal.

On October 9 the attack continued along the whole front, eight tanks of the 4th Battalion coming into action east of Premont and the 17th Armoured Car Battalion, under orders of the Cavalry Corps, operating around Maurois and Honnechy. Two days later, on the 11th, five tanks of the 5th Battalion operated with the 6th Division north of Riguerval wood; this was the last tank action fought in this battle.

The Battle of Cambrai—St. Quentin was at an end. The Hindenburg Line had now to all intents and purposes ceased to exist as an obstacle. It had been broken on a front of nearly thirty miles, on which frontage a penetration of some twenty miles had been effected, and no fewer than 630 guns and 48,000 prisoners captured during the last fourteen days. The effect of this great battle, coupled with the successes of the French in the south and the operations east of Ypres and round Courtrai, fought by the British, French, and Belgians in the north, resulted in the withdrawal of the German forces in the Roubaix, Lille, and Douai area, and with this withdrawal the whole of the British forces in France from north of Menin to Bohain, seven miles north-west of Guise, were faced with field warfare; open country stretched before them, uncut by trench, unhung by wire. The period of exploitation had arrived—that period all our endeavours had been concentrated on attaining during four years of the most desperate and relentless war in history.

Considering the comparative weakness of the British Army, the time of the year, and the nature of the fighting, it had truly been a notable performance on the part of the English and the Dominion infantry, to have fought their way so far. To carry out a rapid pursuit was beyond their endeavours, for the German Army, though beaten, was not yet broken. For cavalry to do so was unthinkable, for the German rear-guards possessed many thousands of machine guns, and as long as these weapons existed, pursuit, as cavalry dream it to be, is utterly impossible. One arm alone could have turned the present defeat into a rout—the tank, but few of these remained, for since August 8 no fewer than 819 machines had been handed over to Salvage by the tank battalions, and these battalions themselves had lost in personnel 550 officers and 2,557 other ranks, a small number indeed when compared with the number of actions the Corps had been engaged in, yet a severe loss out of a fighting state of some 1,500 officers and 8,000 other ranks.

Had it been possible at this crisis to put into the field two fresh brigades of medium tanks, that is about 300 machines, the cost of which would be approximately £1,500,000, or one-fifth that of one day's cost of the war, the greatest war in all history might have closed on or near the field of Waterloo in a decisive victory ending in an unconditional surrender or an irretrievable rout.

Chapter 36
The U.S.A. Tank Corps

On April 2, 1917, the United States of America entered the Great War. Up to this date tanks had not accomplished much. British machines had taken part in the battles of the Somme and Ancre, and the first French ones had made their appearance on the training ground in October 1916.

In June 1917, Lieutenant-Colonel H. Parker was detailed to inquire into the military value of tanks, and in the following month he forwarded his report on this subject to the Operation Section of the Infantry Committee of Colonel C. B. Baker's Commission.

Lieutenant-Colonel Parker's report makes good reading; not only is it virile but sound. It was indeed a great pity that it was not more completely acted on. The following is an extract from it:

> 1. A hole 30 k. wide *punched* through *the whole German formation* deep enough to uncover a line of communication to a flank attack.

This hole must be wide enough to assure the passage of lighter equipment—the divisional machine-gun companies can follow the tanks because the tanks will *make* a road for them.

The wave of machine guns—divisional companies—must turn out to right and left, supported by a second line of tanks, to widen the breach.

The wave of machine guns must be followed by *cavalry*—'hell for leather'—if the hole is once punched through, and this cavalry must strike lines of communication *at all hazards*. Possibly motor-cycle machine guns may be better adapted to this use than cavalry, but *I am a believer in the cavalry*. Support it with Jitney-carried infantry and machine guns as quickly as possible.

2. The problem is *that of passing a defile. Nothing more.* It is like trying to force a mountain pass, where the sides are occupied by enemy who can fire down into the pass. The 'pass' is some 30 k. in length, and we must have something that can *drive through.* Then turn to the sides and widen the breach. Assail 100 k. to cover assault.

It is the old 'flying wedge' of football, with interference coming through the hole in the line. The 'tanks' take the place of the 'line buckers' who open the hole; the 'Divisional Jitney machine guns' are the 'interference,' the 'cavalry' will carry the ball as soon as the hole is opened, *i.e.* ride through and hit the line of communication.

3. The operation works out this way:

(*a*) A cloud of fighting *avions* at high altitude, to clear the air.

(*b*) A cloud of observation *avions* at low altitude, just in front of the line of tanks, dropping bombs and using machine guns on the trenches.

(*c*) Our long-range artillery blocking the German artillery.

(*d*) Our lighter artillery barraging the front to prevent escape of the Germans in their front lines.

(*e*) Our mobile machine guns following up the tanks at about 500 yards, covering them with *canopy fire*, step by step.

(*f*) Our Divisional Jitney companies of machine guns driving in 'hell-bent' after the tanks and widening the breach.

(*g*) Our cavalry riding through this breach as soon as it is opened for them and swinging out *à la* Jeb. Stuart around McClellan's Army. Sacrificed? of course, but winning results worth the sacrifice.

(*h*) Jitney or truck-transported infantry following as fast as gasoline can carry it to support the success and make our foothold sure.

(*i*) Truck-transported—or tank-transported—artillery following as fast as possible.

I believe such a plan will win. Fritz has not the resources to adopt such a plan. We have. We should do it and do it now as far as preparation goes in material. It will take time to get ready.

Shortly before this report was written, Colonel Rockenbach, the commander designate of the American Tank Corps, landed in France and proceeded with General Pershing to Chaumont, the U.S.A. General Headquarters.

On September 23, 1917, a project for a Tank Corps was approved. The Corps was to consist of 5 heavy and 20 light battalions, together with headquarter units, depots and workshops, while in the United States a training centre comprising 2 heavy and 5 light battalions was to be maintained. In May 1918 the establishment of the Corps was expanded to 15 brigades, each brigade to consist of 1 heavy and 2 light battalions, the former to be armed with the Mark VIII and the latter with the Renault tank.

Meanwhile an immense constructional programme was developed for both Mark VIII.s and Renaults, yet, in spite of this, by November 11, 1918, one year and seven months after America entered the war, only some twenty odd American-built Renault tanks had been landed in France. The slowness in American construction is very apparent when it is remembered that a similar period only elapsed between the first sketch drawing of the British Mark I tank, in February 1915, and the landing of this machine in France in August 1916.

The lack of machines in the American Tank Corps rendered the training of its personnel impossible, consequently at the beginning of 1918 two training camps were started, one at Bovington—the British Tank Training Centre—and the other at Bourg in the Haute-Marne, where training was carried out under French supervision. The history of the units trained at these two centres will be dealt with separately as follows:

By February 1918, 500 volunteers from various branches of the American Army were assembled at Bourg for instruction. On March 27, 10 Renault machines were taken over from the French, another

15 being sent to Bourg in June. In August, 144 Renault tanks arrived, and 2 light battalions were at once mobilised under the command of Colonel G. S. Patten and were railed to the St. Mihiel area, where they operated with the First American Army, which attacked the famous salient on September 12.

From a tank point of view this attack was a disappointing one. From railhead both battalions moved 20 kilometres to their positions of assembly, but on the first day of the attack, owing to the difficulties of ground in a well-established defence area, they never succeeded in catching up with the infantry. These troops moved forward rapidly, for it must be remembered that the enemy's resistance was very feeble, the salient having already been partially evacuated by the enemy. Owing to lack of petrol the tanks did not participate in the second day's fighting, and on the third they appear only on one occasion to have come into contact with the enemy and to have collected a number of prisoners. The following day these two battalions were withdrawn practically intact, only three machines being left behind damaged or broken down.

The American tanks next appear fighting side by side with French tank units in the Argonne operations. Profiting by their previous experience, although infantry and tanks had never met on the training ground, the two American tank battalions materially assisted their infantry.

On the first day of the Argonne attack, September 26, it had been intended to keep a reserve of tanks in hand for the second day's operations, but owing to the infantry being held up these went into the attack about noon.

From this date until October 13 these battalions were continually placed at the disposal of the infantry commanders, but were not often called upon to take an active part in operations. Frequently they were moved many miles, to the detriment of their tracks and engines and without achieving any great result; they were also used independently for reconnaissance work and for unsupported attacks delivered against positions the infantry had failed to capture.

On October 13 the remains of these two battalions were withdrawn and a provisional company was formed which accompanied the advance of the American forces until the cessation of hostilities on November 11, 1918.

The 301st U.S.A. Heavy Tank Battalion arrived at Wool on April 10, and continued training under British instruction until August 24, when it embarked for France. Soon after its arrival in this country it was attached to the 1st British Tank Brigade.

On September 29 the 301st American Tank Battalion took part in the important attack carried out by the 27th and 30th American Divisions against the Hindenburg Line running east of the Bellicourt tunnel. The attack started at 5.50 a.m. in a thick mist, and though the 30th American Division reached the Bellicourt tunnel to time, the 27th on its left was held up. On the front of the last-named Division only one tank succeeded in crossing the tunnel, the others running foul of an old British minefield as described in Chapter 35. Of the thirty-four tanks which took part in this attack only ten rallied.

On October 8, when the Fourth Army resumed the offensive, the 301st Battalion was allotted to the IInd American Corps, which was attacking a position some 3,000 yards north-west of Brancourt with the IXth British Corps on its right and the XIIIth on its left. This attack was a complete success; the 301st Battalion fought right through to its final objective, rendering the greatest assistance to the infantry, who worked in close co-operation with the tanks. One tank in particular did great execution: it advanced, firing both its 6-pounders at the railway cutting between Beaurevoir and Montbrehain, the ground being littered with German dead.

Nine days later, on the 17th, the attack was continued, the 301st Battalion again being attached to the IInd American Corps, the objective of which was a line running west of Busigny—eastern edge of La Sablière wood (south of Busigny)—west of Bohain. In this operation the crossing of the River Selle, south of St. Souplet, was a most difficult problem, as the river ran through "No Man's Land"; nevertheless, by means of low-flying aeroplanes reconnaissance and night-patrol work was carried out, crossings were selected, and on the actual day of the attack no fewer than nineteen tanks out of the twenty operating successfully crossed the stream.

The next and last attack carried out by the 301st Battalion during the war took place on October 23, when nine tanks of this unit assisted the 6th and 1st British Divisions in an attack in the neighbourhood of Bazuel, south-east of Le Cateau. This operation was part of the Fourth Army's attack, the objectives of which were the high ground overlooking the canal de la Sambre et Oise, between Catillon, and Bois l'Evêque and the villages of Fontaine-au-Bois, Robersart, and Bousies.

All nine tanks moved forward at zero hour behind the barrage, and from the report of an observer who saw these machines in action it appears that they cleared up the whole of the ground as far as the Bazuel-Catillon road. Very little opposition was met with, but in spite of this, owing to the poor visibility and the enclosed nature of the country, the infantry were slow in following the tanks and great difficulty was experienced in maintaining touch with them. Nevertheless, all infantry commanders expressed themselves well pleased with the work the tanks had accomplished, which had chiefly consisted in reducing strong points and breaking paths through the hedges. Of the nine tanks which took the field all rallied; no casualties other than five men, slightly gassed, were suffered. The attack on this day was altogether a fitting conclusion to the brief but conspicuously gallant career of the 301st American Tank Battalion.

Chapter 37
The Battles of the Selle and Maubeuge

On October 12, the 3rd, 4th, 5th, 7th, and 15th Battalions were withdrawn and placed in G.H.Q. reserve, and on the following day the 6th Battalion was transferred to the 4th Tank Brigade; meanwhile the retiring enemy endeavoured to form a defensive line on the east side of the River Selle.

On this front, on October 17, the Fourth Army and the First French Army attacked from Le Cateau southwards to Vaux Andigny on a front of about twelve miles. The 4th Tank Brigade was the only brigade in action, and its battalions were allotted as follows:

1st Tank Battalion	To the IXth Corps, on the right
301st American Battalion	To the IInd American Corps, in the centre
16th Tank Battalion	To the XIIIth Corps, on the left

The 6th Tank Battalion was held in Fourth Army reserve.

The chief obstacle was the River Selle, the course of which roughly approximated to the starting line in "No Man's Land," consequently reconnaissance of this obstacle was extremely difficult. In spite of this tapes were laid across the stream at night time, when it was discovered that the river had been dammed in places in order to render the crossing of tanks over it more difficult.

The early morning of the 17th was so foggy that tanks had to move forward at zero hour (5.30 a.m.) on compass bearings. Each of the forty-eight machines used carried a crib, and by casting these into the Selle north of St. Souplet and at Molain the 1st and 16th Battalions and the 301st American Battalion crossed this river safely. The resistance offered by the enemy was not great, the Germans apparently having considered the flooded river a certain obstacle against tanks.

Three days later the Third Army attacked between Le Cateau and the Scheldt canal, four tanks of the 11th Battalion co-operating with the Vth Corps against Neuvilly and Amervalles. Again, the chief difficulty was the crossing of the Selle: this was successfully effected by means of an underwater sleeper bridge constructed by the Royal Engineers at night time: being under water the bridge was not visible to the enemy during the day. The attack was entirely successful, all four tanks crossing the river and reaching their objectives.

About the middle of October, the 2nd Tank Brigade was reconstituted, the following battalions being allotted to it: 6th, 9th, 10th, 14th Battalions, and the 301st American Tank Battalion. All these units were short of men and very short of machines.

On October 23, thirty-seven tanks took part in a successful moonlight attack at 1.20 a.m. carried out by the Third and Fourth Armies north and south of Le Cateau, with the object of securing the whole line from the Sambre along the edge of the forest of Mormal to the vicinity of Valenciennes. In this attack the following battalions took part:

301st American Tank Battalion, allotted to IXth Corps.
10th Tank Battalion, allotted to XIIIth Corps.
11th and 12th Tank Battalions, allotted to Vth Corps.

In spite of the darkness, mist, and a considerable amount of gas shelling, all objectives were reached. Many good targets presented themselves, especially for case-shot fire, and in all some 3,000 prisoners were captured. In this attack tanks were of considerable help in crushing down hedges and so opening gaps in them for the infantry to pass through.

The attack was continued on the following day, six machines of the 10th Battalion co-operating with the 18th and 25th Divisions in the neighbourhood of Robersart. Near Renuart farm great assistance was rendered to the infantry, and a German ammunition dump exploded by a 6-pounder shell threw the enemy into great confusion and inflicted many casualties on him.

With this attack the Battle of the Selle came to an end and with it 475 guns and 20,000 prisoners were added to those already captured.

The Battle of Maubeuge opened on November 2 with an attack

carried out by the IXth Corps west of Landrecies. This attack was supported by three tanks of the 10th Battalion and it was carried out in order to improve our position near Happegarbes preliminary to a big attack on the 4th. All objectives were taken, but unfortunately lost again before nightfall.

November 4 witnessed the last large tank attack of the war, large only in comparison with the number of machines at this time fit for action. The attack was on a broad front of over thirty miles, extending from the River Oise to north of Valenciennes. On the British section of this front thirty-seven tanks were used and were allotted as follows:

Third Army	VIth Corps	1 Company 6th Tank Battalion.
	IVth Corps	2 Sections 14th Tank Battalion.
	Vth Corps	1 Company 9th Tank Battalion.
Fourth Army	XIIIth Corps	5 Sections 14th Tank Battalion.
		2 Companies 9th Tank Battalion.
		2 Sections 14th Tank Battalion.
		17th Armoured Car Battalion.
	IXth Corps	4 Sections 10th Tank Battalion.

From the above distribution of tanks, it will be seen how exhausted units had become, sections now taking the place of companies and companies of battalions.

Zero hour varied on the Corps fronts from 5.30 to 6.15 a.m. Briefly the action of the tanks was as follows:

Those of the 10th Tank Battalion assisted in the taking of Catillon and Happeharbes; the capture of the former village was an important step in securing the crossing over the Oise canal. Generally speaking, the tanks operating with the XIIIth Corps had a successful day, especially in the neighbourhood of Hecq, Preux, and the north-western edge of the forest of Mormal. Although supply tanks, (a supply tank is armed with one Lewis gun), are not meant for fighting purposes, three, which were carrying forward bridging material for the 25th Division, came into action near Landrecies.

On approaching the canal, they found that our infantry were still on its western side, hung up by machine-gun fire. One tank being knocked out, the section commander decided to push on with the other two; this he did, our infantry following these machines as if they were fighting tanks, with the result that the machine-gunners surrendered and the far bank of the canal was secured.

The following day, November 5, saw the last tank action of the war, eight Whippets of the 6th Tank Battalion taking part in an attack of the 3rd Guards Brigade north of the forest of Mormal. The country was most difficult for combined operations, for it was intersected by numerous ditches and fences which rendered it ideal for the rearguard operations the Germans were now fighting all along their front. Either the Whippets had to go forward and so lose touch with our infantry or remain with the infantry and lose touch with the enemy. In spite of these difficulties all objectives were taken, and the last tank action of the war was a success.

During the next few days refitting continued with a view to building up an organised fighting force from the shattered remnants of the Tank Corps; as this was in progress the signing of the armistice terms on November 11 brought hostilities to an end.

Ninety-six days of almost continuous battle had now taken place since the great tank attack at Amiens was launched by the Fourth Army on August 8, since when many of the officers and men of the Tank Corps had been in action as many as fifteen and sixteen times. During this period no fewer than 1,993 tanks and tank armoured cars had been engaged on thirty-nine days in all; 887 machines had been handed over to Salvage, 313 of these being sent to the Central Workshops, and 204 having been repaired and reissued to battalions.

Of the above 887 tanks, only fifteen had struck off the strength as unsalvable. Casualties against establishment had been heavy: 598 officers and 2,826 other ranks being counted amongst killed, wounded, missing, and prisoners; but when it is considered that the total strength of the Tank Corps on August 7 was considerably under that of an infantry division, and that in the old days of the artillery battles, such as the First Battle of the Somme, an infantry division frequently sustained 4,000 casualties in twelve hours fighting, the tank casualties were extraordinarily light. It was no longer a matter of twelve hours' but of thirty-nine days' fighting at twelve hours a day. From this we may deduce our final and outstanding lesson from all these battles, namely, that iron mechanically moved is an economiser of blood, that the tank is an economiser of life—the lives of men, men being the most valuable asset any country can possess.

The determination of Sir Douglas Haig had at length been rewarded, and the endeavours which failed at Passchendaele won through finally and irrevocably at Maubeuge. A fitting conclusion to all these operations is to be found in the last dispatch of the Commander-in-Chief of the British Armies, which hands down to posterity a just judgment on the value of the work carried out by the British Tank

Corps during the ever-memorable months of August to November 1918. In these dispatches are to be found the following three paragraphs, which are worth pondering over when the time comes for us to consider the future:

> In the decisive contests of this period, the strongest and most vital parts of the enemy's front were attacked by the British, his lateral communications were cut and his best divisions fought to a standstill. On the different battle fronts 187,000 prisoners and 2,850 guns were captured by us, bringing the total of our prisoners for the present year to over 201,000. Immense numbers of machine guns and trench mortars were taken also, the figures of those actually counted exceeding 29,000 machine guns and some 3,000 trench mortars. These results were achieved by 59 fighting British divisions, which in the course of three months of battle engaged and defeated 99 separate divisions.
>
> Since the opening of our offensive on August 8, tanks have been employed on every battlefield, and the importance of the part played by them in breaking up the resistance of the German infantry can scarcely be exaggerated. The whole scheme of the attack of August 8 was dependent upon tanks, and ever since that date on numberless occasions the success of our infantry has been powerfully assisted or confirmed by their timely arrival. So great has been the effect produced upon the German infantry by the appearance of British tanks that in more than one instance, when for various reasons real tanks were not available in sufficient numbers, valuable results have been obtained by the use of dummy tanks painted on frames of wood and canvas. It is no disparagement of the courage of our infantry, or of the skill and devotion of our artillery, to say that the achievements of those essential arms would have fallen short of the full measure of success achieved by our armies had it not been for the very gallant and devoted work of the Tank Corps, under the command of Major-General H. J. Elles.

CHAPTER 38

The 17th Tank Armoured Car Battalion

In March 1918 the 17th Tank Battalion was in process of formation at the Tank Training Centre at Wool, when the German spring offensive resulted in so great a demand being made on the home resources that it was converted into an Armoured Car Battalion on April 23. On the following day the drivers were selected, and sixteen armoured cars, which were earmarked for the eastern theatre of war, were handed over to it, the Vickers machine guns being replaced by Hotchkiss ones.

On April 28 the cars were embarked at Portsmouth, and on the 29th the personnel, under the command of Lieutenant-Colonel E. J. Carter, left Folkestone for Boulogne. Thus, in six days the whole battalion was formed, equipped, and landed in France.

Immediately on landing the 17th Battalion was attached to the Second Army and ordered to proceed to Poperinghe, but the tactical situation improving these orders were cancelled and it was first sent to the Tank Gunnery School at Merlimont for instruction, and later on to the Tank Depot at Mers.

After some ten days' training the 17th Battalion joined the Fourth Army and went into the line at La Hussoye, being attached to the Australian Corps. A few days later the battalion was transferred to the XXIInd Corps, which was then resting in G.H.Q. reserve, immediately behind the right flank of the British Army, and battalion headquarters were established at Pissy. Here training continued until June 10, when at 9.30 a.m. instructions were received by Lieutenant-Colonel Carter to report to the headquarters of the First French Army at Conty.

At Conty orders were issued for the battalion to proceed to Ravenel near St. Just. The battalion was notified of this by telephone, and, although the night was very dark and wet and the roads crowded with traffic, it reached Ravenel by 5 a.m. on June 11, after a sixty-mile journey, and went into action with the Tenth French Army in its counter-attack at Belloy on that day. In this battle two sections of armoured cars engaged the enemy with machine-gun fire, but the quantity of debris scattered on the roads, and the fragile nature of the chassis of the cars, prevented their being freely used. On the conclusion of these operations the battalion returned to the XXIInd Corps.

On July 18 the 17th Battalion was ordered to join General Fayolle's Army, but did not come into action on account of delays on the road due to congested traffic. Ten days later it was attached to the 6th French Cavalry Division, which was operating north of Château-Thierry following the retreating enemy towards the River Ourcq. On this river the Germans took up a defensive position, covering its approaches by machine-gun fire; this brought the French cavalry to a halt, but not the armoured cars, which were able, on account of their armour, to approach quite close to the bridges and open fire on the enemy's machine-gunners. At Fère-en-Tardenois the battalion greatly aided the French by moving through the main streets of the village, which was held by Germans. Similar assistance was rendered to the Americans at Ronchères.

When the 6th French Cavalry Division was withdrawn to rest the 17th Battalion proceeded to Senlis, and at 9 a.m., having just entered this town, it received orders to proceed forthwith to Amiens, and report to the headquarters of the Australian Corps. Amiens, which was nearly 100 miles distant, was reached the same night.

On arriving at Amiens Lieutenant-Colonel Carter was informed that his unit was to take part in the projected attack east of this town. The chief difficulty foreseen in an armoured-car action in this neighbourhood was the crossing of the trenches. Although only one day was available wherein to find a solution to this difficulty, it was accomplished by attaching a small force of tanks to the battalion. These tanks were used to tow the armoured cars over the obstacles, or rather, along the tracks the tanks formed through them. This solution proved eminently successful.

For the Fourth Army operations the 17th Armoured Car Battalion was placed under the orders of the 5th Tank Brigade. On the morning of August 8, the battalion moved forward with its accompanying tanks, which successfully assisted all its cars over "No Man's Land." Beyond Warfusée, several large trees, felled by shell fire, had fallen across the road, entirely blocking it; these were speedily removed by the towing tanks, thus clearing the road not only for the armoured cars but for our guns and transport. After this delay the cars moved rapidly forward and passed through our attacking lines about twenty minutes before the infantry were timed to reach their final objective. To accomplish this the cars had to run through our own artillery barrage; this they did without casualty.

The road was now clear and the cars proceeded through the enemy's lines, scattering any infantry they found on the road. They made for the valley near Foucaucourt, where the headquarter troops of a German Corps were known to be encamped. These troops were completely surprised and many casualties were inflicted on them by six cars moving through the valley. The confusion caused soon developed into a panic, the enemy scattering in all directions, spreading the alarm.

Whilst this surprise was developing, several sections of armoured cars turned south and north off the Amiens-Brie road. The former met large columns of transport and mounted officers and teams of horses apparently belonging to the German headquarters at Framerville. These were fired on at short range, four officers being shot down by a single burst of fire. Shortly after this the German headquarters were reached, and the Australian Corps flag, which had been carried in one of the cars for the purpose, was run up over the house which, until a few minutes before, had been occupied by the German Corps Commander. At about this time one car came in sight of a German train: the engine was fired at and put out of action; later on the cavalry arriving captured it.

The cars which had turned northwards entered Proyart and Chuignolles, two moving up to the River Somme. At Proyart the cars found the German troops at dinner; these they shot down and scattered in all directions, and then, moving westwards, met masses of the enemy driven from their trenches by the Australians. In order to surprise these men, who were moving eastwards, the cars hid in the outskirts of Proyart, and, when the enemy was between fifty and one hundred yards distant, they rapidly moved forward, shooting down great numbers. Scattering from before the cars at Proyart the enemy made across country towards Chuignolles, only to be met by the cars which had proceeded to this village, and were once again fired on and dispersed. Near Chuignolles one armoured car obtained "running practice" with its machine guns at a lorry full of troops, and kept up fire until the lorry ran into the ditch. There were also several cases of armoured cars following German transport vehicles, without anything unusual being suspected, until fire was opened at point-blank range.

Although more than half the cars were out of action by the evening of the 8th there were no casualties amongst their personnel sufficiently serious to require evacuation.

After repairing the damages sustained on August 8, the 17th Battalion was transferred to the First Army, and on August 21 took part in the operations near Bucquoy. At the entrance of the village a large crater had been blown in the road, over which the cars were hauled after a smooth path had been beaten down across it by a Whippet tank. The cars then made their way through the enemy's lines and reached Achiet-le-Petit ahead of our infantry, where several machine guns were silenced by them. In this action two of the cars received direct hits, one of them being burnt out and destroyed.

On August 24 the battalion operated with the New Zealand Division in the attack on Bapaume, the cars penetrating to the Arras-Bapaume road, where severe fighting took place.

In the attack of September 2, the 17th Battalion operated with the Canadian Corps in the assault on the Drocourt-Queant line. In this action four cars were hit by shell fire, but two squadrons of aeroplanes co-operating with the cars attacked the German battery so vigorously that the crews of the disabled cars were able to escape being captured.

On September 29 the armoured cars operated with the Australian Corps and the IInd American Corps in the attack on the Hindenburg Line near Bony; here numerous casualties were inflicted on the enemy and four cars were put out of action by being burnt. This position was captured by the Australians on the following day.

On October 8 the armoured cars were attached to the Cavalry Corps, which was operating from Beaurevoir towards Le Cateau. On this day the cars kept touch with the cavalry, but on the following morning they moved forward through Maretz. About two miles beyond this village a section co-operated with South African infantry and drove the German machine-gunners from a strong position they were holding. The cars were able to run right through the hostile machine-gun fire, and by enfilading the enemy's position killed the German machine-gunners and captured ten machine guns and two trench mortars.

A section of cars made a dash to cross the railway bridge on the Maretz-Honnechy road, but the enemy's demolition party saw them coming and, lighting the fuse, fled. The leading car, however, got across safely, the charge exploding and blowing up the bridge immediately this car had crossed and thereby cutting it off from the second car, which was some fifty yards behind. The leading car then went through Maurois and Honnechy, all guns firing; both of these villages were crowded with troops. Near Honnechy church the car ran into a by-road by mistake; at the same moment a group of Germans came out of a house and the car accounted for five of them in the doorway. This incident was described with enthusiasm by a French woman, the owner of the house, to Lieutenant-Colonel Carter on the following day. After passing Honnechy the car was run towards a bridge which was known to exist.

Profiting by his previous experience the commander of the car determined to save the bridge from demolition and so not only effect his retreat but secure it to the British Army. To accomplish this the car rapidly moved round a corner of the road leading to the bridge, with its guns pointing in the direction where the demolition party would probably be. This action proved successful; the demolition party being scattered by a burst of bullets before the charge could be fired. The bridge was thus saved and proved of great importance to the British forces later on. The car then crossed the river and proceeded to the spot where the second car had been unable to cross, picking it up; both cars returned to report their action, one at least having accomplished a very daring and useful journey.

On November 4 the armoured cars were attached to the XVIIIth Corps and were detailed to operate with the 18th and 50th Divisions in the forest of Mormal. In this district the roads are narrow and at this time of the year were very slippery; armoured-car action was therefore most difficult. On the next day the cars of the 17th Battalion, now much reduced in numbers, were operating with no fewer than five divisions simultaneously. On the 9th all cars were concentrated and attached to the Fourth Army advanced guard to assist in the pursuit of the retiring enemy. In the action which followed the cars were cut off from the advanced guard by all the river bridges being destroyed, but in spite of this they were able to continue advancing on a line parallel to the pursuit. At Ramousies and Liessies three complete trains of ammunition were passed and numbers of heavy guns, lorries and artillery transport, the enemy being in full flight and in a high state of demoralisation.

On November 11 the armoured cars were reconnoitring towards Eppé-Sauvage and Moustier (twelve miles east of Avegnes), near the Belgian frontier, some seven or eight miles in advance of the nearest British troops, when at 10.30 a.m. an officer from the 33rd French Division informed the officer in command that he had heard rumours of an armistice; a few minutes later a dispatch-rider corroborated this information, stating that hostilities were to cease at 11 a.m. Firing went on until about three minutes to eleven, when it ceased, breaking out in a final crash at eleven o'clock—then all was silence; a silence almost uncanny to the men of the 17th Tank Armoured Car Battalion, who had not been out of gunshot since July 17, the date upon which the battalion opened its eventful history with the French Army on the Marne.

Dramatic as had been the short and brilliant career of the 17th Armoured Car Battalion, its work was not yet ended. On November 13 it assembled at Avesnes, and joining the cavalry of the Fourth Army moved forward towards the Rhine. On the 26th four sections of cars were ordered to Charleroi to deal with a reported disturbance. In this town they were received with the greatest enthusiasm by the inhabitants, and at Courcelles were surrounded by excited townsfolk who, having collected all available brass instruments, crowded round the cars playing the British National Anthem at a range of about five yards.

From Charleroi, the 17th Battalion joined the Second Army, moving on Cologne, and were attached to the 1st Cavalry Division. On December 1 the German frontier was crossed at Malmédy, whence the battalion was immediately sent on with the 2nd Cavalry Brigade to deal with disturbances which had broken out in Cologne. Five days later, on the 6th, the cavalry halted outside the town, and the G.O.C. 2nd Cavalry Brigade, escorted by cars of the 17th Battalion, proceeded to the Rathaus to discuss the administration of the town with the burgomaster. Cologne was entered at midday, the crews of the armoured cars being the first British troops to enter. That afternoon the western end of the Rhine bridge was occupied, and the colours of the Tank Corps run up to flutter over the famous river.

The record of this battalion is a truly remarkable one. It was formed, equipped, and landed in France in the short space of six days. In six months, it fought in ten separate battles with English, Australian, Canadian, New Zealand, South African, French, and American troops, and was three times mentioned in German dispatches. Every car was hit and some of them many times, and yet the total losses in killed in action throughout this period was only one officer and four other ranks. At the cost of these five men and seven cars totally destroyed, this battalion must have inflicted scores if not hundreds of casualties on the enemy. That the British Army was not equipped with many more of these units will be a problem which will doubtless perplex the minds of future military historians.

Chapter 39

A Retrospect of What Tanks Have Accomplished

Like all other human energies, war may be reduced to a science, and had this, throughout history, been better understood, how many countless thousands of lives and millions of money might not have been saved, and how much sorrow and waste might not have been prevented!

Science is but another name for knowledge—knowledge co-ordinated, arranged and systematised—from which art, or the application of knowledge to existing and ever-changing conditions, is derived and built up on unchanging principles.

The fundamental difficulty in the art of war is in the application of its theories in order to test their values. Like surgery and medicine, it demands its patients or victims as its training-ground, and without these it is most difficult to arrive at expert judgments and conclusions. It is an art which is neither directly commercial, materially remunerative, nor normally applicable, consequently it has generally been looked upon as a necessary evil, an insurance against disaster rather than the application of a science which should have as its main object the prevention of the calamity of war.

As an applied science, war is half human, half mechanical; it is, therefore, pre-eminently a live or dynamic science, a science which must grow with human understanding itself, so that its means of action, materialised in the soldier, may not only keep level with progress but absorb it to its own particular ends. When we look back on the history of war, what do we see? A school of pedants fumbling with the past, hoodwinked against the future, seeking panaceas in past victories, the circumstances under which these were won being blindly accepted as recurring decimals. Thus, do they lumber their minds with obsolete detail, formulæ and shibboleths, precepts and rituals which are as much out of place on the modern battlefield as phlogiston or the philosopher's stone would be in a present-day laboratory.

Time and again has it been asserted that war itself is the sole test of a soldier's worth and that on the battlefield alone will the great be sifted from the little.

And why? Because, until today, we have never emerged from what may be called the "alchemical" epoch of warfare, the compounding of illusions without knowledge, the application of actions without understanding; we have not reduced war to a science founded on definite principles nor learnt that 99 *per cent.* of victory depends on weapons, machinery placed in the hands of man so that he may kill without being injured.

Galen was a great physician and so was Paracelsus, but who today would apply their methods when they can employ those of Pasteur and Lister? Where we have been so wrong and will continue to remain so wrong, unless we radically change our peace methods of warfare, is that we possess no process of producing great peace soldiers—scientists for war. We do not realise that an army is formed to prevent war, that it is composed of human points, that the good player will not lose many of these points, and that the bad player will go bankrupt.

That the loss or gain depends on superiority of brains and of weapons and not necessarily superiority of rank and numbers of men. When we do realise this, then shall we cast the ancient balsams, solvents, and coagulants to the winds and set about developing the mental and mechanical sides of war in days of peace, so that, should wars become inevitable, we may win them with the minimum of human loss.

Soldiers have laughed at Joly de Maizeroy, Massenbach, and Maurice de Saxe for suggesting "victory without fighting," "wars without battles"; but seldom are their eyes dimmed with a tear when they read of a victory which cost thousands of lives, and a victory which might have been won at the cost of a few hundreds. Yet surely is the saving of men's lives and limbs as great an attribute of good leadership as the taking of those of the enemy; is it not in fact endurance, or the staying power in human lives, which is the backbone of victory itself?

In August 1914 the Great War opened to all intents and purposes as an exaggerated 1870 operation. The doctrine of the contending armies was 1870, its leaders were saturated with 1870 ideas, its weapons were improved 1870, it was 1870 in complexion, in tone, in manner, in thought, in tactics, and in movement. If this be doubted read the text-books prior to the war and compare them with those of 1872 and then with the events of the war itself. Take any great army of 1918 and place it over the same army of 1914: the sides do not coincide. What is the one great difference? Mechanical progress in weapons, not numbers of men, for men potentially had in numbers decreased; yet any 1918 equipped army would have beaten a 1914 one because of guns, heavy guns, super-heavy guns, mortars, shells, bombs, grenades, gas, machine guns, machine rifles, automatic rifles, range-spotters, sound-detectors, smoke, aeroplanes, lorries, railways, tramways, armoured cars, and tanks.

What today would be thought of a mechanical engineer who applied 1870 methods? Nothing; he would go bankrupt in six weeks if he started business on 1870 lines. This is exactly what the armies of 1914 did; they tactically went bankrupt because they were sufficiently big, or the area of operations was sufficiently small, to deny to them strategical movement. Could this have been foreseen? Given the numbers, given the weapons, and given the area of operations, a simple rule-of-three sum can be worked out, the answer to which is siege warfare and the tactics of which is the frontal attack of penetration, (see "The Tactics of Penetration," by Captain J. F. C. Fuller, *Journal of the Royal United Services Institution*, November 1914, this article was written in April 1914); yet every *Field Service Regulations,* in 1914, favoured envelopment and paid but a passing attention to trench warfare.

Inevitably the preordained tactics of penetration were forced on the contending parties, and human points were thrown over the parapets in handfuls; as if men, armed with a rifle and bayonet, who could only secure their existence by remaining underground, had any chance whatever of attaining a decisive victory by forsaking their shelters and facing weapons in the open which had previously forced them to earth. What was the result? The Germans failed at Ypres and Verdun; the French in the Champagne, at Verdun, and at Reims; and we at Neuve Chapelle, Loos, the Somme, Arras, and Passchendaele.

Between 2,000,000 and 3,000,000 casualties on one side of the balance sheet and a few square miles of uninhabitable ground on the other was the sum-total of these united endeavours, and all because no single army had, since 1870, realised the mechanical side of the science of war. In October, ten weeks after the war had opened, as the second chapter of this book has already related, the mechanical side was realised and a solution was found in the production of a chariot not so very dissimilar to that depicted on the "Victory Stele of Eannatum" of Lagash, 3,000 years B.C.—no very novel mechanical invention!

Time, a few months, was, however, requisite for the substitution of the petrol engine for the horses of the Assyrians, and as time could not be wasted other mechanical lapses were made good which might have well been foreseen had penetration and not envelopment been diagnosed as the leading tactical act of the war.

At first each contending nation in turn passed through its barbed-wire crisis, its gun shortage and its ammunition scandal. Millions of miles of wire were produced, thousands of guns were made, and ammunition was manufactured not by thousands of rounds, but by hundreds of thousands of tons. Had any one side been able to fire at the other, in September 1914, 100,000 tons in a couple of days, that side would have, probably, won the war. This is practically what happened at the Dunajec in 1915—the Russians were out-weaponed and consequently defeated.

On the Western Front, as the artillery competition was more or less mutual, stagnation became still more complete. In place of hurling men against uncut wire, shells were hurled instead, the bombardments being sufficiently long to enable the Germans to transport troops from the east of Poland to France in time to meet the assault. As the frontage of this assault was usually under ten miles, the total battle-front being over 500, the operation may be compared with that of attempting to take the life of a rhinoceros with a hat-pin. These tactics inevitably failed, not only through the impossibility of economically wearing away the enemy's reserves, but on account of the impossibility of rapidly moving forward our own; for in the act of destroying wire, simultaneously did the guns create an area so difficult to move over that, had it been possible to advance the infantry, it would never have been possible to feed or supply them.

That stationary warfare should have increased in endurance as the gun-power of each side was multiplied was not necessary; this was clearly proved during the first two German battles of 1918. By this date, on all sides, had artillery attained its zenith, but the Germans, by threatening a front of nearly 250 miles—practically from the Channel to the Meuse—and then, after an intense bombardment lasting but a few hours, attacking on a comparatively wide front of some fifty miles, were able to develop their machine power to its fullest effect, that is to say, with the least opposition.

It took nearly three years from the date of the Battle of the Dunajec before the use of the gun as a weapon of surprise was grasped; this will probably prove one of the most astounding tactical anomalies of the war. During this period two other weapons were devised which were destined in most respects to outclass the gun; the idea of both must have arisen at approximately the same time.

For years before the war the French and ourselves had been the leading mechanical engineers of Europe; in a similar respect the Germans were its leading chemists. Both, once a deadlock had arisen in the war, sought aid from the sciences they best understood during peacetime, and from which, had they understood war as a science, they would have looked for assistance years before its present outbreak.

The first stroke of genius delivered in the war was the use the Germans made of gas on April 22, 1915, and the second the use we made of tanks on September 15, 1916; both failed through want of a scientific grasp of war. They were tentative attacks, not delivered in strength or mass, yet curious to relate both were delivered by armies which, having been brought up in the 1870 school of thought, were fully conversant with the old precept of "superiority of numbers at the decisive point"; but, thinking in muscular terms only, they failed to apply it to the mechanical and chemical contrivances now placed at their disposal.

By many soldiers even today it is not realised that gas is a missile weapon following directly along the evolutionary path of all projectiles. A solid shot has to hit a target in order to injure it; as targets became difficult to see it became necessary to increase the radius of effect of the solid shot by replacing it by a hollow one filled with explosive. By means of this shell, a target might be missed by the shell yet hit by a flying fragment; the danger zone of the solid shot was increased many hundreds of times. Once targets not only become invisible but disappear into under-earth shelters, the shell has but little effect unless days are spent in bombardment, consequently the most effective manner of hitting them is to replace the shell by a gas inundation which will cover extensive areas and percolate into trenches and shelters. Gas has, in fact, multiplied the explosive radius of action of a shell indefinitely, and had it been used in quantity by the Germans before the Allies could protect themselves against it, the enemy might well have won the war.

Gas, whatever its possibilities were before this protection was obtained, remains but a projectile evolved as above described. Tanks were a "creation," and the introduction of the petrol-driven cross-country tractor on the battlefield, it is thought, will mark a definite close to the "alchemical" epoch of warfare. All war on land, in the past, has been based on muscular energy; henceforth it will be based on mechanical. The change is radical, and Wilson's "Big Willie" will one day pass into legend alongside Stevenson's "Rocket." As steam, applied as a motive force, in 150 years changed the world more than it had previously been changed since the days of Paleolithic man, so, before the present century has run its course, may as great a change take place in the realms of war.

The cause of both is the same: as the invention of the steam engine rendered obsolete to a high degree the hand-tool and replaced it by the machine-tool, so the application of petrol to the battlefield will force the hand-weapon out of existence and replace it by the machine-weapon. That the tank will continue in its present form is as unlikely as it would have been to expect, in 1769, that Watt's pumping engine was the "Ultima Thule" of all such engines. It is not the form which is the stroke of true genius, but the idea, the replacing of muscular energy by mechanical force as the motive power of an army.

Had the combatant nations of the Great War possessed more foresight, had they thought of war as a science in place of as an insurance policy, they could have had a steam-driven tank thirty years ago and a petrol-driven one immediately after the South African War. The Batter tractor existed, anyhow in design, in 1888, and during the South African War Mr. W. Ralston drew a comic picture entitled "Warfare of the Future: The Tractor Mounted Infantry in Action," to say nothing about the story by Mr. H. G. Wells. But no, the breath of ancient battles had to be breathed, and whilst military students were studying Jena, Inkerman, and Worth, the commercial sciences were daily producing one invention after another which a little adjustment would help win the next war more speedily than the study of scores of Jominis and Clausewitzs.

To show how unscientific the soldiers of the 1870–1914 epoch had become it is only necessary to quote that after the Battle of the Somme in the highest German military circles the machine was considered as a veritable joke. Apparently, it could not be seen that, though the Mark I tank was far from perfect, it, being able to reintroduce armour and to provide the soldier with a mobile weapon platform, revolutionised the entire theory of 1870 warfare.

On July 1, 1916, the opening day of the Battle of the Somme, the British Army sustained between 40,000 and 50,000 casualties. On September 25, one single tank forced the surrender of 370 Germans at a cost of five casualties to ourselves, yet in July 1917 the Mark IV tank was still considered but as a minor factor. Its design was not sufficiently reliable, its true powers were more or less a matter of conjecture; the troops were not fully accustomed to it, nor would they place sufficient faith in it to accept it in lieu of artillery support, in fact, in its present state of development the tank was but an adjunct to infantry and guns. Such were some of the views held regarding it when, like a bolt from the blue, the Battle of Cambrai shot across the horizon of 1870 battles.

At Cambrai it was the Mark IV tank which was used, the same which had existed in July; the Tank Corps had not increased materially in size; the infantry were for the most part used-up troops—some had received a few days' training with tanks, others had never even seen these machines. The assault was an overwhelming success: at the cost of some 5,000 infantry casualties an advance was made in twelve hours which in extent took ninety days at Ypres, and which in this last-named battle cost over a quarter of a million men. Yet, in spite of this astonishing success, so conservative had the army grown to the true needs of victory that there were certain soldiers who now stated that the tactics employed at Cambrai could never be repeated again and that the day of the tank had come and gone.

★★★★★★

Then came the "crowning mercy"—the attack on Hamel. Something had to be done to reinstate the credit of the Tank Corps. There were but three suitable localities to do it in: the first, against the Merville salient—the ground here was bad, being intersected by dikes and canals; the second, eastward from between Arras and Hebuterne—the ground here was much cut up, and the tactical objective was not suitable; the third, eastwards from Villers-Bretonneux—the ground here was excellent, but the Australians, who held this sector, had little confidence in the tank.

Human prejudice is, however, not difficult to overcome to the student of psychology. After tactful persuasion the Australian Corps was induced to accept sixty machines, as an "adjunct" to their operations. The tanks (Mark V.s) were drawn up 1,000 yards *in rear* of the attackers, yet, nevertheless, within a few minutes of the attack being launched, they caught up with the leading wave and carried this wave and those in rear right through to the final objective. The loss of the 4th Australian Division was insignificant; their prejudices vanished and a close comradeship between them and the Tank Corps was established which redounds to their gallantry and common sense.

Hamel, minor incident though it was, was of more importance to the immediate problems of the British Army than Cambrai itself. General Rawlinson, commanding the Fourth Army, saw his opportunity, and the result was that from Hamel onwards the war became a tank war. The machine had made good in spite of prejudice and opposition. The Germans lost their heads, and with their heads they lost the war. That the war might have been won without tanks is quite possible, but that fifty-nine British divisions would, without their assistance, have beaten ninety-nine German ones in three months is extremely unlikely.

What had the influence of the tank really been? Let us examine this question and so close this retrospect.

The effect of the tank's mobility on grand tactics was stupendous. Between the winter of 1914 and the summer of 1918, to all intents and purposes, the Allies waged a static war on the Western Front. During these three and a half years various attempts were made to wear down the enemy's fighting strength as a prelude to a decisive exploitation or pursuit, but these battles of attrition were mutually destructive and the Allies undoubtedly lost more casualties than they inflicted. Attrition without the possibility of surprise or mobility is a mere "push of pikes," it is a muscular but brainless operation. At the Third Battle of Ypres it cost us a quarter of a million men. Then came the tank, and true attrition was rendered possible; in other words, in tank battles the enemy lost more in human points than we did: it is doubtful whether in killed and wounded we lost, between August 8 and November 11, 1918, as many men as the prisoners we captured. This was only possible by our possessing the means of putting the grand tactical act of penetration into operation, by breaking down the "inviolability" of the German front, and by so doing rendering envelopment a reality.

In minor tactics it was possible, by means of the tank, to economise life by harmonising fire and movement and movement and security; the tank soldier could use the whole of his energy in the manipulation of his weapons and none in the effort of moving himself forward; further than this, sufficiently thick armour could be carried to protect him against bullets, shrapnel, and shell splinters. Human legs no longer controlled marches, and human skin no longer was the sole protection to the flesh beneath it. A new direction was obtained, that of the moving firing line; the knight in armour was once again reinstated, his horse now a petrol engine and his lance a machine gun.

Strategy, or the science of making the most of time for warlike ends, had practically ceased since November 1914. Even the great advances of the Germans in 1918 came to an abrupt stop through failure of road capacity, and roads and rails form the network upon which all former strategy was woven. The cross-country tractor, or tank, widened the size of roads to an almost unlimited degree. The earth became a universal vehicle of motion, like the sea, and to those sides which relied on tanks, naval tactics could be superimposed on those of land warfare.

With the introduction of mechanical movement every principle of war became easy of application and, today, to pit an overland mechanical army against one relying on roads, rails, and muscular energy, is to pit a fleet of modern battleships against one of wind-driven three-deckers. The result of such an action is not even within the possibility of doubt: the latter will for a certainty be destroyed, for the highest form of machinery must win, because it saves time, and time is the controlling factor on the battlefield as in the workshop.

CHAPTER 40
A Forecast of What Tanks May Do

Accepting war as a science and an art, that it is founded on definite principles which are applied according to the conditions of the moment, we may scientifically reduce it to its component elements,

GENERAL

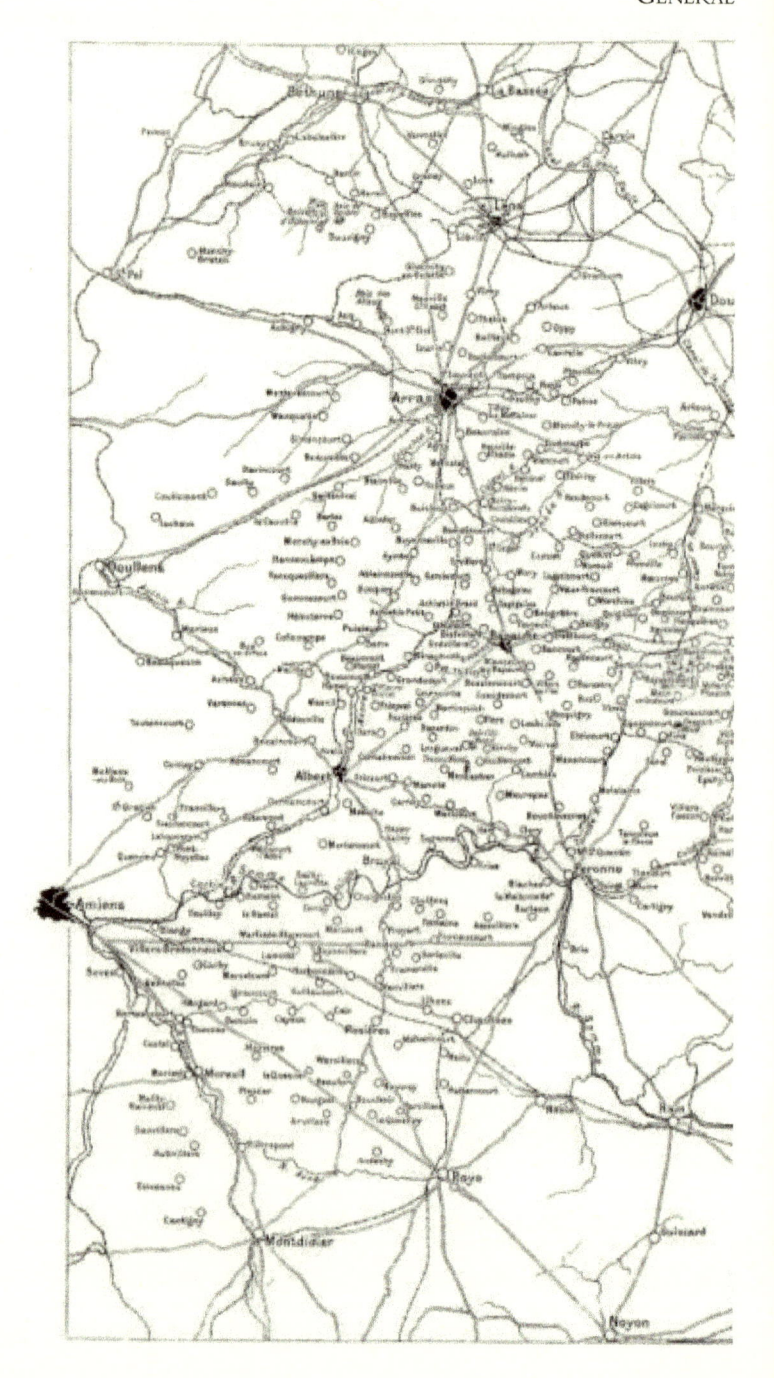

MAP

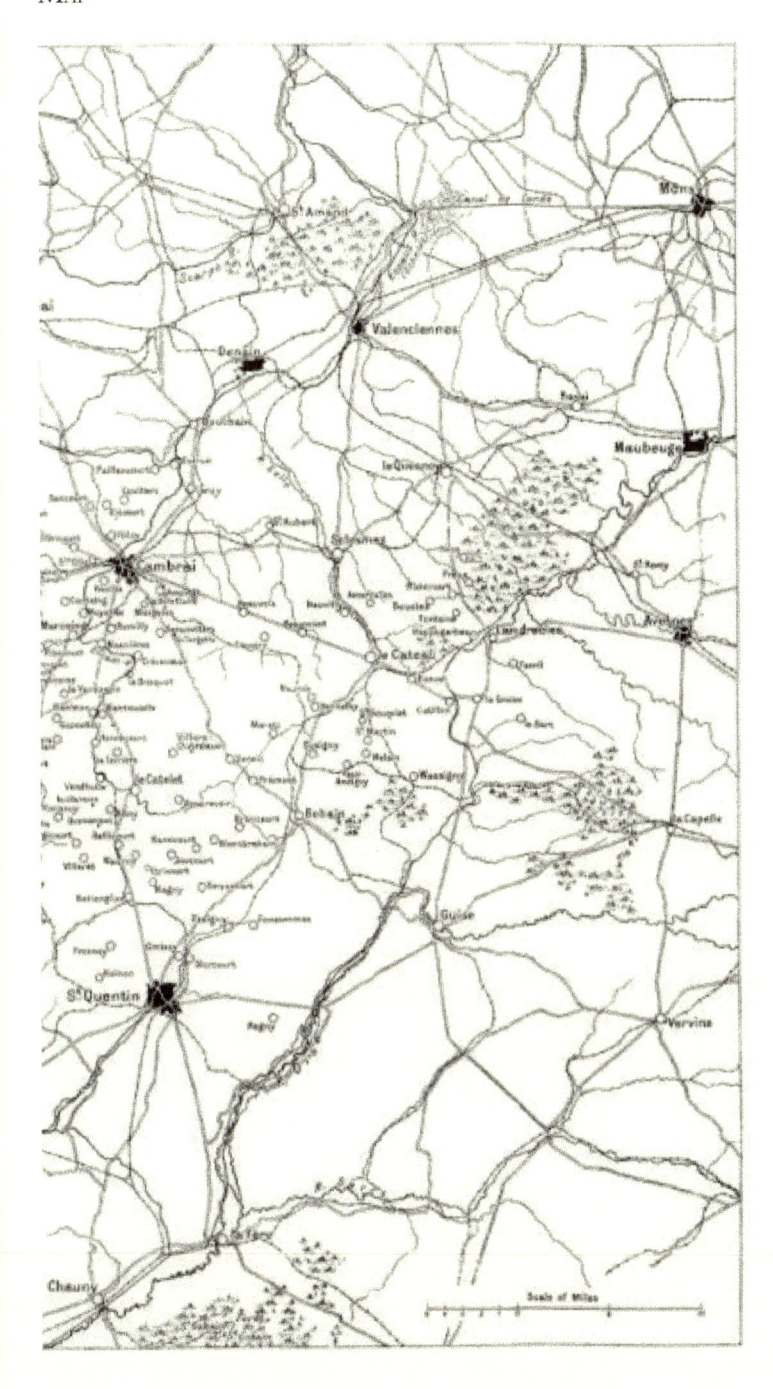

which are: Men, weapons, and movement. A combination of these three is an army, a body of men which can fight and move.

Tactics, or the art of moving armed men on the battlefield, change directly in accordance with the nature of the weapons themselves and the mobility of the means of transport. Each new or improved weapon or method of movement demands a corresponding change in the art of war.

Tools, or weapons, if only the right ones can be discovered, form 99 *per cent.* of victory. Strategy, command, leadership, courage, discipline, supply, organisation, and all the moral and physical paraphernalia of war are as nothing to a high superiority of weapons; at most they go to form the 1 *per cent.* which makes the whole possible. Indeed, as Carlyle writes, "Savage animalism is nothing, inventive spiritualism is all."

Today the introduction of the tank on the battlefield entirely revolutionises the art of war in that:

(1) It increases mobility by replacing muscular force by mechanical power.

(2) It increases security by rendering innocuous the effect of bullets through the feasibility of carrying armour plate.

(3) It increases offensive power by relieving the man from carrying his weapons or the horse from dragging them, and by facilitating ammunition supply it increases the destructive power of the weapons it carries.

In other words, an army moved by petrol can obtain a greater effect from its weapons in a given time with less loss to itself than one which relies on muscular energy as its motive force. Whilst securing its crew dynamically a tank enables it to fight statically, it is in every respect the "landship" it was first called.

These are our premises and from them we may deduce the following all-important fact: That in all wars, and especially modern wars—wars in which weapons change rapidly—no army of fifty years before any date selected would stand a "dog's chance" against the army existing at this date, not even if it were composed entirely of Winkelrieds and Marshal Neys. Consider the following examples:

(1) Napoleon was an infinitely greater general than Lord Raglan; yet Lord Raglan would, in 1855, have beaten any army Napoleon, in 1805, could have led against him, because Lord Raglan's men were armed with the Minie rifle.

(2) Eleven years after Inkerman, Moltke would have beaten Lord

Raglan's army hollow, not because he was a greater soldier than Lord Raglan, but because his men were armed with the needle gun.

From this we may deduce the fact, which has already been stated, namely, that weapons form 99 *per cent.* of victory, consequently the General Staff of every army should be composed of mechanical clairvoyants, seers of new conditions, new fields of war to exploit, and new tools to assist in this exploitation. Had Napoleon, in 1805, offered a prize of £1,000,000 for a weapon 100 *per cent.* more efficient than the "Brown Bess," it is almost a certainty that, by 1815, he would have got it; for the want of a little foresight and for the want of the understanding that progress in weapons of war is a similar problem to progress of tools in manufacture, he might have saved his Empire and ended his days as supreme tyrant of Europe.

The whole history of the evolution of machine tools is that of the elimination of the workman and the replacement of muscular energy by steam, electricity, or some other form of power. "*Fewer men, more machines, higher output*" has during the last hundred years been the motto of every progressive workshop. Likewise, we believe that from now onwards in every progressive army will a similar motto be adopted. Further than this, we believe that those nations which have proved their ability in the past as leaders of science and mechanical engineering will in the future be those which will produce the most efficient armies, for these armies will be based on the foundations of the commercial sciences.

Accepting that the main factor in future warfare will be the replacing of man-power by machine-power, the logical deduction is that the ideal army to aim at is *one* man, not a conscripted nation, not even a super-scientist, but one man who can press a button or pull a plug and so put into operation war-machines evolved by the best brains of the nation during peacetime. Such an army need not even occupy the theatre of operations in which the war is to be fought; *he* may be ensconced thousands of miles away, perhaps in Kamtchatka, fighting a battle on the Western Front. Is this impossible? Not at all; even in the late war we can picture to ourselves a one-armed cripple sitting in Muravieff-Amourski and electrically discharging gas against the Hindenburg Line directly his indicator announces a favourable wind.

So far, the chemist, but is man going to be controlled by gas, are human destinies to be limited by a "whiff of phosgene"?

"Certainly not," answers the soldier mechanic. "It is true that the future may produce many unknown gases which, as long as they remain unknown to the opposing side, are unlikely to be rendered innocuous by means of a respirator; I, however, will scrap the respirator and place my men in gas-proof tanks, and whenever my indicator denotes impure air, the crews will batten down their hatches, their engines will be run off accumulators, and they themselves will live on oxygen or compressed air. I will apply to land-warfare naval methods undreamt of before, I will produce a land machine which will, so to speak, submerge itself when the gas cloud approaches, just as a submarine submerges in the sea when a destroyer draws near."

There is an answer to every weapon, and that side which has most thoroughly thought these answers out during days of peace is the one which is most likely to produce a steel-shod Achilles for days of war.

Without journeying so far as Amourski let us imagine that war was to break out again three years hence and that we were equipped with a tank 200 *per cent.* superior to our at present best type—a machine travelling at fifteen miles an hour in place of five, and that the Germans sitting behind their Hindenburg Line were still backing personnel against *matériel*, numbers of men against perfection of weapons.

An army is an organisation, comparable, like all other organisations, very closely with the human body. It possesses a body and a brain; its fighting troops are the former, its headquarters staffs the latter. In the past the usual process of tactics has been to wage a body warfare: one body is moved up against the other body and like two boxers they pummel each other until one is knocked out. But suppose that boxer "A" could by some simple operation paralyse the brain of boxer "B," what use would all boxer "B's" muscular strength be to him, even if it rivalled that of Samson and Goliath combined? No use at all, as David proved!

Now apply this process to the battle of 1923. The tank fleets, under cover of dense clouds of smoke, or at night-time, move forward, not against the body of the enemy's army but against his brains; their objectives are not the enemy's infantry or the enemy's guns, not positions or tactical localities, but the billets of the German headquarters staffs—the Army, Corps, and Divisional headquarters. These they capture, destroy or disperse; what then is the body going to do, for its brain is paralysed? Who is going to control it, feed it with reserves, ammunition, and supplies? Who is going to manoeuvre it to give it foot play? Either it will stand still and be knocked out, or, much more likely, it will be seized by panic and become paralysed to action.

What is the answer to this type of brain warfare? The answer is the tank; the brains will get into metal skulls or boxes, the bodies will get into the same, and land fleet will manoeuvre against land fleet.

The growth of these tactics may be slow, but eventually they will become imperative. It may be urged that the field gun is master of the tank in the open, just as a land battery is master of a ship at sea. This is only true as long as the gunner can see his target, and no known means at present exist whereby sight can penetrate a dense cloud of smoke. It may also be urged that a heavy machine gun will enable the infantry to protect themselves against tanks. But to be mobile the weight of the machine gun is limited to the carrying power of two men—about 80 lb., and there is no known reason why a tank should not be armoured to withstand the bullets of such a weapon. If a heavier machine gun is made it will be forced to take to a mounting, and for choice to a mechanical one; it will in fact become a tank or a tank destroyer.

The necessity of armour in war has always been recognised, and its general disuse only dates from the sixteenth century onwards. When armour could not be used other means of protection, all makeshifts, were sought after—earth-works, entrenchments, use of ground, manoeuvre, and covering fire, and as regards the last-named substitute it is interesting to go back a little into history, for, even from a cursory study, we may better understand the present and foresee the future.

In the days of our Henry VIII a body of arquebusiers had to stand twenty-five ranks deep in order to obtain continuity of fire; that is to say, that once the first rank had fired and doubled to the rear it would only be ready loaded again when the twenty-fifth rank was about to discharge its pieces. By the days of Gustavus Adolphus, the art of musketry and the musket had so far improved as to permit of these twenty-five ranks being reduced to eight. As improvement went apace we find Frederick the Great reducing them to three, and Wellington in the Peninsula to two. Even in the early period of the revolutionary wars it was found necessary for light infantry to reduce the human target they offered to the enemy's fire by making use of extensions.

In 1866 extensions became more feasible on account of the Prussians being armed with a breach-loading rifle; in 1870 they became more general; in 1899 they have grown to between ten and fifteen paces, which may be taken as the maximum for a man, armed with the magazine rifle, to deliver one round per yard of front each minute. In 1904 trenches are made use of on an extensive scale, for as extensions cannot be increased if fire effect is to be maintained, some other form of protection must be sought, and men, not being able to carry armour, must carry spades instead and so still further immobilise themselves. In 1914, after a brief hurry-scurry of open warfare, all sides take to earth and the spade reigns supreme.

Then comes the reintroduction of armour with the tank, and what do we see? Not only mobility and direct protection, but the reinstitution of the firing line, not now morcellated at fifteen paces interval between the men composing it, but at 150 to 300 paces between the tanks, the mechanical skirmishing fortresses of which it is built up. A tank with a crew of 6 men can deliver fire at the rate of 300 rounds a minute, or equivalent to 30 riflemen at a South African War extension, and being armoured they suffer practically no loss and can consequently challenge not only 30 riflemen but 300, any number, in fact, who are sent against them. The logical conclusion to be drawn from this is that extensions are useless, trenches at best but static makeshifts, the infantryman must don armour and, as he has not the strength to carry it, he must get into a tank. If this is common sense, let us attempt to visualise what a tank war of the future may entail.

In the mechanical wars of the future we must first of all recognise the fact that the earth is a solid sea as easily traversable in all directions by a tractor as a sheet of ice is by a skater; the battles in these wars will therefore more and more approximate to naval actions. As trenches, as we know them, and the ordinary field obstacles now constructed will be useless, it may become necessary during peacetime to turn the great strategical centres—manufactories, railways, stores, seats of government, etc., into defended land-ports or protected power, fuel, and control stations. The fortifications of these will probably consist of immense dry moats and extensive minefields which will constitute a direct protection against tank attacks. Water obstacles will be useless, for the tank of a few years hence will undoubtedly be of an amphibious nature. To protect these centres from the air, barracks, storehouses, mobilisation stores, tankodromes and aerodromes will all have to be constructed well beneath the surface of the ground—in fact, the future fortress will approximate closely to a gigantic dugout surrounded by a field of land mines electrically manipulated.

Near the frontier these defended ports will probably be equipped with and linked up by lethal gas works—gas-producing and storage plants, lodged below the surface, which on war being declared can instantaneously be set operating electrically by one man stationed hundreds of miles away if needs be. When this type of warfare is instituted, mobilisation will not consist in equipping with weapons a small section of the community, but in providing such of the civil population as cannot be rapidly evacuated from the area it is proposed to inundate, or placed in gastight shelters safely underground, with anti-gas appliances. Under these circumstances the defence of frontiers will be organised according to prevailing winds, and signs of war will be looked for not amongst military but civil movements.

As the gas-storage tanks are opened and the gas-producing plants set operating, fleets of fast-moving tanks, equipped with tons of liquid gas, against which the enemy will probably have no means of protection, will cross the frontier and obliterate every living thing in the fields and farms, the villages and cities of the enemy's country.

✶✶✶✶✶✶

During the war the normal system of detecting new gases was to examine captured respirators, and from the chemicals they contained inversely deduce the gases they would protect their wearers against. In peacetime no such means of detection will be possible.

✶✶✶✶✶✶

Whilst life is being swept away around the frontier fleets of aeroplanes will attack the enemy's great industrial and governing centres. All these attacks will be made, at first, not against the enemy's army, which will be mobilising underground, but against the civil population in order to compel it to accept the will of the attacker.

If the enemy will not accept peace terms forthwith, then wars in the air and on the earth will take place between machines to gain superiority. Tank will meet tank, and, commanded from the air, fleets of these machines will manoeuvre between the defended ports seeking each other out and exterminating each other in orthodox naval fashion. Whilst these small forces of men, representing perhaps 0·5 *per cent.* or 1 *per cent.* of the entire population of the country, strong through machinery, are at death-grips with their enemy, their respective nations will be producing weapons for them; so, in the future, as military fighting man-power dwindles must we expect to see military manufacturing man-power increase.

Are we safe in this little island of ours against the future? If at times, during the "alchemical" period of warfare, we have been threatened and invaded, we may be certain that during the scientific period we shall be less secure than we have been in the past.

From the present-day tank to one which can plunge into the

Channel at Calais at 4 in the morning, land at Dover at six o'clock, and be outside Buckingham Palace for an early lunch will not probably require as many as the fifty-two years which have separated the *Merrimac* from the *Tiger* or the *Queen Elizabeth*. If this is too remote a period for the present generation to grow anxious about, there is no reason why four or five years hence ships should not be constructed as tank-carriers, these machines being conveyed across the ocean and launched into the sea near the coast carrying sufficient fuel to move them 300 or 400 miles inland. From ships as carriers it is but one step to aeroplanes as suppliers and lifters, and another to aeroplanes as tanks themselves.

If the evolution of war, in the past, has been slow, do not let us flatter ourselves that it is likely to remain so in the future. From the gliders of the Wright Brothers the aeroplane rapidly evolved, and from a 40 H.P. engine of ten or twelve years back today the Porte "Super-Baby" triplane carries five engines of 400 H.P. each and the Tarrant triplane has a span of 131 feet and to drive it six Napier "Lion" engines are used, developing no less than 3,000 horse-power. The tank is still in its infancy, but it will grow and one day in mechanical perfection and efficiency catch up with the super-Dreadnought and the Handley Page, and what then? A close co-operation between the great mechanical weapons, the seaship, the airship, and the landship—or, if preferred, of boat, aeroplane, and tank—will take place. These weapons will approximate and unify, evolving one arm and not three arms, which will require one defence force and not three. This, even today, is becoming more and more apparent, and the sooner the brains of the future Defence Force are developed the better for this nation, for today we are thinking, like medieval magicians, in separate terms of air, water, and earth, and some of us in those of gabions, lances, and blunderbusses.

If great wars can be restricted or abolished by word of mouth or written agreement, the above gropings into the future, even if possible, may never materialise; but even if this be so, many small wars lie in front of us, for Europe politically, since 1914, has practically gone back 400 years, the frontiers of the smaller nations approximating closely to those of the later Middle Ages. The more nations there are in the world the more wars there will be in the making, and as half the smaller nations of central and eastern Europe consider war a national sport there is little likelihood of agreements being kept or peace being maintained; in fact, all agreements which cannot be compelled by brute force are likely to be treated as "scraps of paper."

To enforce peace, power and the means of applying it will be needed by the greater nations who by law will never quarrel; here the mechanic steps forward and presents the nations concerned with the tank and the aeroplane as a means towards this end. He is perfectly right; the general introduction of mechanical weapons must bring with it the end of small wars if not also of civil disturbances.

Take the case of the defence of India. What has always been the great difficulty in our frontier expeditions? Not our enemy or his weapons, but the country which enables the Afghan to evade our columns and impede our advance. It is the resistance offered by natural obstacles which we have to overcome and not those imposed upon us by weapons which generally are vastly inferior to those with which our men are equipped.

Take the case of a punitive expedition starting from Peshawar and proceeding to Kabul. The force will consist of three bodies of troops—a small fighting advanced guard, a large main body protecting the transport, and strong flank guards protecting the main body. On account of the tactics which have to be adopted the advance is excessively slow. The main body proceeding along the roads, which almost inevitably coincide with the bottoms of the valleys, has to be kept out of rifle shot, consequently the flank guards have usually to "crown the heights" on each side of the road, which necessitates much climbing and loss of time. If the advance were over an open veldt land, as in South Africa, in place of in a hilly country, movement would be simplified, but still will the flank guards have to be thrown out because the main body, consisting of men and animals, is pervious to bullets. This perviousness to bullets is the basis of the whole trouble, and unless bullet-proof armour can be carried, when it does not matter whether the rifle is fired at a range of two yards or two miles, the only means of denying effect to the rifle is to keep it out of range of its target.

Though up to a short time ago the carrying of armour was not a feasible proposition, now it is, and there are few more difficulties in advancing up or down the Khyber with a well-constructed tank than across the open. Armour, by rendering flesh impervious to bullets, does away with the necessity of flank guards and long straggling supply columns, and our punitive expedition equipped with tanks can reach Kabul in a few days, and not only reach it but abandon its communications, as they will require no protection. If tank supply columns, which are self-protecting, are considered too slow, once the force has reached Kabul its supply and the evacuation of its sick (there will be but few wounded) can be carried out by aeroplane. The whole operation becomes too simple to be classed as an operation of war. Once impress upon the Afghan the hopelessness of facing a mechanical punitive force and he will give up rendering such forces necessary.

In our many small wars of the past we have frequently been faced with desert warfare, a warfare even more difficult than hill and mountain fighting. Here again the chief difficulty is a natural one—want of water, and not an artificial one—superiority of the enemy's weapons. In 1885 Sir Henry Stewart started from Korti on the Nile to relieve General Gordon: his difficulties were supply difficulties, and it took him twenty-one days to reach Gubat, a distance of 180 miles. A tank moving at an average pace of ten miles an hour could have accomplished the journey in two days, and being supplied by aeroplane could have reached Khartum a few days later. One tank would have won Maiwand, Isandhlwana, and El Teb; one tank can meet any quantity of Tower muskets, or Mauser rifles for aught that; one tank, costing say £10,000, can not only win a small war normally costing £2,000,000, but render such wars in the future highly improbable if not impossible. The moral, therefore, is—get the tank.

From small wars to internal Imperial Defence is but one step. Render rebellion hopeless and it will not take place. In India we lock up in an unremunerative army 75,000 British troops and 150,000 Indian. Both these forces can be done away with and order maintained, and maintained with certainty, by a mechanical police force of 20,000 to 25,000 men.

What now is the great lesson to be learnt from the above examples? That war will be eliminated by weapons, not by words or treaties or leagues of nations; by weapons—leagues of tanks, aeroplanes, and submarines—which will render opposition hopeless or retribution so terrible that nations will think not once or twice but many times before going to war. If the civilian population of a country know that should they demand war they may be killed in a few minutes by the tens of thousands, they will not only cease to demand it but see beforehand that they are well prepared by superiority of weapons to terrify their neighbours out of declaring war against them.

Weapons we, therefore, see are, if not a means of ending war and ridding the world of this dementia, a means of maintaining peace on a far firmer footing than hitherto it has been maintained by muscular power. To limit the evolution of weapons is therefore to limit the periods of peace. An army cannot stand still, it must develop with the civilisation of which it forms part or become barbaric. To equip our army today with bows and arrows would not reduce the frequency of war, it would actually increase it, for according to his tools, so is man himself, and as an army is built up of men, if these men are armed with bows and arrows they will in nature closely approximate to the age which produced these weapons, the age which burnt Joan of Arc. Equally so will the army of today, if in equipment it be not allowed to keep pace with scientific progress, develop into a band of brigands, for in 2019 the rifle and gun of today, and the civilisation which produced them, will be as uncouth as the arquebus, the carronade, and the manners of the sixteenth century.

If a millennium is ever to be ushered in upon earth it will be accomplished through the development of brain-power and not through it becoming atrophied. If war is to be rendered impossible the process will be a slow evolutionary one, the desire of war gradually slowing down, and its motive force energising some other ideal. To restrict war by maintaining soldiers as ill-armed barbarians is to prevent it working out its destined course. Human nature, in spite of Benjamin Kidd, does not change in a generation, and the tendencies which beget war will out until human nature has outgrown them. The world has a soul, and like that of a man it must pass through years of love, hate, striving and ambition before attaining those of wisdom and decay.

There may yet be many wars ahead of us, but one thing would appear to be certain, and this is that small wars will disappear and great ones become less frequent, science rendering them too terrible to be entered upon lightly.

Today we stand upon the threshold of a new epoch in the history of the world—war based on petrol, the natural sequent of an industry based on steam. That we have attained the final step on the evolutionary ladder of war is most unlikely, for mechanical and chemical weapons may disappear and be replaced by others still more terrible. Electricity has scarcely yet been touched upon and it is not impossible that mechanical warfare will be replaced by one of a wireless nature, and that not only the elements, but man's flesh and bones, will be controlled by the "fluid" which today we do not even understand. This method of imposing the will of one man on another may in its turn be replaced by a purely psychological warfare, wherein weapons are not even used or battlefields sought or loss of life or limb aimed at; but, in place, the corruption of the human reason, the dimming of the human intellect, and the disintegration of the moral and spiritual life of one nation by the influence of the will of another is accomplished.

Be all these as they may, one fact stands out supreme in all types and conditions of war, and this is, that the strongest and most efficient brain wins, which applies equally to all nations as it does to all individuals.

Animal superiority over animal is based on muscle, human superiority over human is based on brain. The nation with the supreme brain will eventually rule the world, and so long as war continues the Army with the best brains (which also means the best weapons) will accomplish victory with the least loss. Our army from today must step forward; "to advance is to conquer," and this applies in greater force to brain-power than to muscle-power, for brains control muscles. To stand still is to retrogress; to glance backwards is to lose time, and if we pause now, we are lost in the future.

Do not, therefore, let us mark time on our own graves, do not let us hark back to 1914 with its rifles and its ammunition boots, its sabres and its horseshoes, and all its muscular barbarism; let us plan and let us think, thus shall we penetrate the veil of the future, thus shall we learn how to equip our army with a brain and with a body which united, if war be ever again forced upon us, will compel victory at the smallest possible cost. Surely this is an ideal worthy of a great nation and of a great army, the object of which is to prevent war and to maintain peace, to prevent war by science and not by nescience, by progress and not by retrogression.

ALSO FROM LEONAUR
AVAILABLE IN SOFTCOVER OR HARDCOVER WITH DUST JACKET

WINGED WARFARE *by William A. Bishop*—The Experiences of a Canadian 'Ace' of the R.F.C. During the First World War.

THE STORY OF THE LAFAYETTE ESCADRILLE *by George Thenault*—A famous fighter squadron in the First World War by its commander..

R.F.C.H.Q. *by Maurice Baring*—The command & organisation of the British Air Force during the First World War in Europe.

SIXTY SQUADRON R.A.F. *by A. J. L. Scott*—On the Western Front During the First World War.

THE STRUGGLE IN THE AIR *by Charles C. Turner*—The Air War Over Europe During the First World War.

WITH THE FLYING SQUADRON *by H. Rosher*—Letters of a Pilot of the Royal Naval Air Service During the First World War.

OVER THE WEST FRONT *by "Spin" & "Contact"* —Two Accounts of British Pilots During the First World War in Europe, Short Flights With the Cloud Cavalry by "Spin" and Cavalry of the Clouds by "Contact".

SKYFIGHTERS OF FRANCE *by Henry Farré*—An account of the French War in the Air during the First World War.

THE HIGH ACES *by Laurence la Tourette Driggs*—French, American, British, Italian & Belgian pilots of the First World War 1914-18.

PLANE TALES OF THE SKIES *by Wilfred Theodore Blake*—The experiences of pilots over the Western Front during the Great War.

IN THE CLOUDS ABOVE BAGHDAD *by J. E. Tennant*—Recollections of the R. F. C. in Mesopotamia during the First World War against the Turks.

THE SPIDER WEB *by P. I. X. (Theodore Douglas Hallam)*—Royal Navy Air Service Flying Boat Operations During the First World War by a Flight Commander

EAGLES OVER THE TRENCHES *by James R. McConnell & William B. Perry*—Two First Hand Accounts of the American Escadrille at War in the Air During World War 1-Flying For France: With the American Escadrille at Verdun and Our Pilots in the Air

KNIGHTS OF THE AIR *by Bennett A. Molter*—An American Pilot's View of the Aerial War of the French Squadrons During the First World War.

AVAILABLE ONLINE AT **www.leonaur.com**
AND FROM ALL GOOD BOOK STORES

ALSO FROM LEONAUR
AVAILABLE IN SOFTCOVER OR HARDCOVER WITH DUST JACKET

OFFICERS & GENTLEMEN by Peter Hawker & William Graham—Two Accounts of British Officers During the Peninsula War: Officer of Light Dragoons by Peter Hawker & Campaign in Portugal and Spain by William Graham.

THE WALCHEREN EXPEDITION by Anonymous—The Experiences of a British Officer of the 81st Regt. During the Campaign in the Low Countries of 1809.

LADIES OF WATERLOO by Charlotte A. Eaton, Magdalene de Lancey & Juana Smith—The Experiences of Three Women During the Campaign of 1815: Waterloo Days by Charlotte A. Eaton, A Week at Waterloo by Magdalene de Lancey & Juana's Story by Juana Smith.

JOURNAL OF AN OFFICER IN THE KING'S GERMAN LEGION by John Frederick Hering—Recollections of Campaigning During the Napoleonic Wars.

JOURNAL OF AN ARMY SURGEON IN THE PENINSULAR WAR by Charles Boutflower—The Recollections of a British Army Medical Man on Campaign During the Napoleonic Wars.

ON CAMPAIGN WITH MOORE AND WELLINGTON by Anthony Hamilton—The Experiences of a Soldier of the 43rd Regiment During the Peninsular War.

THE ROAD TO AUSTERLITZ by R. G. Burton—Napoleon's Campaign of 1805.

SOLDIERS OF NAPOLEON by A. J. Doisy De Villargennes & Arthur Chuquet—The Experiences of the Men of the French First Empire: Under the Eagles by A. J. Doisy De Villargennes & Voices of 1812 by Arthur Chuquet.

INVASION OF FRANCE, 1814 by F. W. O. Maycock—The Final Battles of the Napoleonic First Empire.

LEIPZIG—A CONFLICT OF TITANS by Frederic Shoberl—A Personal Experience of the 'Battle of the Nations' During the Napoleonic Wars, October 14th-19th, 1813.

SLASHERS by Charles Cadell—The Campaigns of the 28th Regiment of Foot During the Napoleonic Wars by a Serving Officer.

BATTLE IMPERIAL by Charles William Vane—The Campaigns in Germany & France for the Defeat of Napoleon 1813-1814.

SWIFT & BOLD by Gibbes Rigaud—The 60th Rifles During the Peninsula War.

AVAILABLE ONLINE AT **www.leonaur.com**
AND FROM ALL GOOD BOOK STORES